The Basics of Permaculture Design

Ross Mars

Artwork by
Martin Ducker

CHELSEA GREEN PUBLISHING COMPANY
White River Junction, Vermont

A Permanent Publications Book

Published by:
Permanent Publications
Hyden House Ltd.
The Sustainability Centre, East Meon, Hampshire GU32 1HR, UK
Tel: (01730) 823 311
Fax: (01730) 823 322
Email: enquiries@permaculture.co.uk
Web: www.permaculture.co.uk

Published in the United States in 2005
by Chelsea Green Publishing Company
www.chelseagreen.com

Published in 2003
by Permanent Publications, UK

First Published 1996
by Candlelight Trust, W.A., Australia

© 1996 Candlelight Trust, W.A., Australia

Artwork by Martin Ducker

Typesetting and layout by Ross Mars

British Library Cataloguing in-Publication Data
A catalogue record for this book is available from the British Library.

ISBN 1 85623 023 6

All rights reserved. Apart from any fair dealing for the purpose of study, research, criticism or review, as permitted under the Copyright Act, no part of this publication may be reproduced, stored in a retrieval system, rebound or transmitted in any form or by any means, electronic, mechanical, photocopying, recording or otherwise, without the prior written permission of Ross Mars, Hyden House Limited or Chelsea Green Publishing Company.

Acknowledgements
A sincere thank you goes to the following people who have commented on various chapters of this book and made valuable contributions to its production:
    Graham Bell, Peter Bennett, Dora Byrne, David Coleman, Naomi Coleman, Peter Cuming, Chris Dixon, Martin Ducker, Calen Feorn, Pratibodha Foreman, David Holmgren, Max Lindegger, Jeff Nugent, Peter Pedals, Salli Ramsden, Margaret Sampey, Pat Scott, Reny Mia Sley, Russell Turner, Patrick Whitefield, Julie Woodman, Peter Woodward.
A special thank you to Barbara Sheppard and Bob Gehringer for permission to use the design of their Donnybrook property for the cover.

Foreword

by David Holmgren

The permaculture concept, developed by Bill Mollison and myself in the mid 70's, is the conscious design of our working relationship with nature. Further development, application and extension to a wider public have resulted in an explosive growth towards a worldwide network of activity and influence.

Many would see this growth as part of a fad and fashion element in the "global village". Others see it as the result of Mollison's personal energy and charisma. While these factors are undoubtedly elements in the growth of permaculture over the last twenty years, in the absence of any organisational structure and management for growth, it is reasonable to suggest that permaculture does provide some of the conceptual links and practical solutions needed to re-establish our working relationship with nature.

It is now widely accepted, almost to the level of cliché, that design is the central organising skill of the post-industrial revolution. That enough basic knowledge exists, but that this knowledge is theoretical and divided into specialist disciplines rather than being practical and integrated, is increasingly understood.

Permaculture involves the integration of ecological design principles, the ethics and values of working with nature and the detailed situation and site-specific practical realities of life. In trying to combine these three very different spheres of human activity, there is constant tension and the need to re-assert balance.

In *The Basics of Permaculture Design* we see the results of one experienced permaculture practitioner, designer and teacher's efforts to maintain this balance. Ross Mars has written a clear and readable guide to permaculture design. The personal experience in dealing with the realities of the affluent society and the West Australian environment come through as hallmarks of practical knowledge, while the design process orientation and frequent examples provide a link to a wider readership.

The mass solutions in land use, livelihoods and language of our industrial culture have failed us. The hope and search for new mass solutions contradicts the site and situation specific characteristics of nature. What we need are universal, powerful and comprehensible design principles for guiding practical and diverse development.

This is the "holy grail" of permaculture which needs ongoing effort and focus. However, in life, we always lack complete understanding and yet, we must act. Ross Mars has provided a useful tool for effective action now.

David Holmgren, Hepburn. August 1996

Contents

Foreword by David Holmgren ... iii
1. Permaculture is a direction, not a destination ... 1
   - what permaculture is and what it is not ... 1
   - ecosystems ... 4
   - sustainable land use ... 7
2. Maximising the edge ... 10
   - land-based food production ... 10
   - water-based food production ... 14
3. General design principles ... 18
   - elements ... 18
   - zones and succession ... 19
   - sectors ... 22
   - microclimate ... 23
     - frost ... 24
     - aspect ... 26
     - reflection and radiation ... 27
   - designing for catastrophe ... 28
     - wind ... 28
     - fire ... 28
     - abundance of water ... 31
     - other considerations ... 31
4. Steps in the design process ... 32
   - the design process ... 32
   - design considerations ... 34
   - design steps ... 35
     - information phase ... 36
     - analysis phase ... 37
     - design phase ... 38
     - management phase ... 41
   - implementing a design ... 41
5. Basic tools for the designer ... 43
   - a designer's field tool kit ... 43
     - tape measure ... 43
     - piece of string ... 43
     - penetrometer ... 44
     - folding shovel ... 44
     - magnetic compass ... 44
     - pH test kit ... 44
     - salinity meter ... 45

	• surveying the landscape	45
	- bunyip level	45
	- "A" frame level	46
	- dumpy level	47
	- more expensive equipment	47
	• a designer's drawing tool kit	48
	- scale ruler	48
	- light box	48
	- stationery items	48
	- mathematical drawing aids	50
	- drawing paper	50
6	Basic principles of garden building and management	51
	• introduction	51
	• building healthy soil	51
	- the nature of soil	51
	- why mulch?	53
	- soil conditioning and treatments	54
	• integrated pest management	56
	• stacking	60
	• guilds	62
	• other tips for gardeners	63
7	It's all a matter of location	65
	• location and climate	65
	• choosing a property	67
	• local regulations	70
8	Getting the house right: zone 0	72
	• the passive solar home	72
	• integrating the house and garden	75
	• retrofitting existing houses	80
9	Water harvesting	84
	• on the surburban block	84
	• on the farm	86
	- keyline cultivation	86
	- dams	89
	- moving water through drains	90
	• in dry areas	96
10	Designs for urban settlement	99
	• gardens	99
	- compost	99
	- garden areas	100
	- ponds	102

- animals ... 104
  - earthworms ... 104
  - bees ... 105
  - poultry ... 106
- limited spaces ... 108
  - balconies and windows ... 108
  - small backyards ... 110
  - germinating seed ... 113

11 Designing for rural properties ... 115
- whole farm plans ... 115
- windbreaks and shelterbelts ... 116
- animals in the system ... 122

12 Permaculture in schools ... 127
- a needs analysis ... 127
- determining resources ... 127
- guidelines for designing school grounds ... 128
- practical design considerations ... 131

13 Communities ... 135
- hamlet - to be or not to be ... 135
- design considerations for ecovillages ... 137
  - bylaws and regulations ... 137
  - guidelines for land development ... 138
  - looking at options ... 139
  - housing structures ... 142
- social and legal structures of human settlements ... 144

14 Appropriate technology ... 148
- generating power ... 148
  - photovoltaic cells ... 148
  - wind generators ... 149
  - hydro-electric systems ... 151
- electric fencing ... 153
- pumping water ... 154
  - solar pump ... 155
  - hydraulic ram ... 155
- cooking devices ... 156
  - solar oven ... 156
  - solar food dryer ... 157
  - haybox cooker ... 157

15 Glossary ... 159
16 Bibliography ... 164
17 Index ... 166

1 Permaculture is a direction, not a destination

What permaculture is and what it is not

Permaculture deals with our existence on this planet and it encompasses many different aspects of this. Firstly, permaculture is about producing edible landscapes, mirroring the natural ecosystems in their diversity and production. Permaculture is primarily a design system. This is the main difference between it and all other agricultural and horticultural practices. Permaculture designs endeavour to integrate all components of the ecosystem in a holistic approach to sustainable living and practice.

Permaculture started out as *perma*nent agri*culture* and thus focussed on the growth and development of perennial food crops. Annuals and biennials do have their place, but the use of long-living food crops, such as fruit and nut trees, is the priority. Some areas of the garden need to be devoted to annuals and in most cases they can be interplanted between the perennial herbs and other trees as companion planted guilds. Too often, annuals are taken for granted in food production and they should be used in the system within the framework of perennial production.

Figure 1.1 A garden should have diversity of annuals and perennials.

Permaculture is not just about gardening, although its origin of permanent agriculture suggests this. Nowadays, permaculture is thought of along the lines of permanent culture, incorporating all aspects of human beings and human settlements.

Gardening, however, is one simple way in which people can take some responsibility for their own existence and begin to care for the Earth. Helping yourself and others to build gardens in your own backyard, in an effort to drastically reduce the need to buy produce from someone else, is one of the most environmentally-responsible things you can do to help reduce our consumption of resources and to heal the planet.

Since the late seventies the concepts of permaculture have also developed, such that it encompasses finances, water harvesting, communities, buildings, and alternative and appropriate technology. For many of us, permaculture is a framework that unites many disciplines, and so the subjects of aquaculture, ethical investment, horticulture, solar technology, soils, and many others can be integrated together, each contributing as part of the whole.

This framework permits many different forms of knowledge to be interwoven - all relative to one another. It is not a set of techniques *per se*, but rather how a number of techniques are employed to build a system in which energy is harvested, directed and allowed to flow, bearing in mind that it is always cheaper to conserve energy than to produce it. Permaculture is also different from both organic gardening and forest gardening in that both of these are techniques of garden construction and composition. Permaculture is more than this. It is a design strategy.

Permaculture is the harmonious integration of design with ecology. The ethics of earth care, people care, limits-aware and surplus-share are common to all permaculturists, even though the design strategies and the techniques they employ vary widely. We design for long term sustainability, and this is why a design is a harmonious integration of landscape, plants, animals and humans, as well as the placement of components or elements in recognisable patterns.

Truly successful designs create a self-managed system. A large amount of what we call permaculture is really just commonsense, using human intuition and insight to solve problems that confront us.

The outcomes of good design should include:

- sustainable land use strategies, without wastes and pollution.
- established systems for healthy food production, and maybe some surplus.
- restoration of degraded landscapes, resulting in conservation of endemic species - especially rare and endangered species.
- integration and harmony of all living things on the property - all things live in an atmosphere of co-operation or interact in natural cycles.
- minimal consumption of energy.

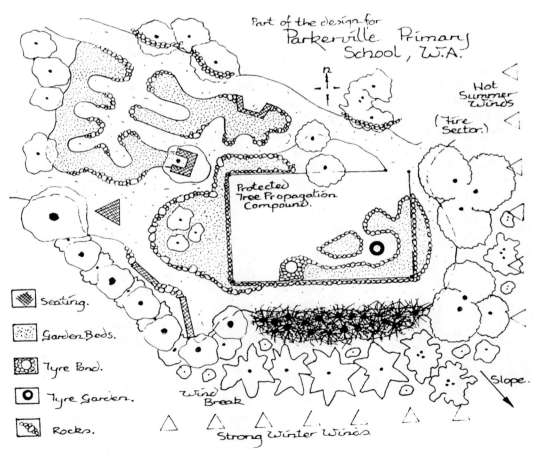

Figure 1.2 A permaculture design is more than a garden plan. It must consider all aspects of the interaction between organisms. (Part of the design of gardens at Parkerville Primary School).

The ultimate design, if there is such a thing, is the marriage of what is best for the land and what is best for the people who live there. What we call a "design" is really only a pictorial representation of the implied inter-relationships between objects, structures, plants, animals and humans. The drawing only gives information about placement and types of species and nothing about their interaction, which is the most important thing about any ecosystem.

For some permaculturists, difficulties arise because of their limited vision, their restricted world view and their world experience. There are some people who may not have had the experience of growing fruit and vegetables, nurturing the soil or building passive solar houses, yet they can come up with a plan for somebody and tell them this is what they need.

In one sense, these armchair permaculturists may be perpetuating myths about how well permaculture works, when they haven't ever fully implemented a design themselves or observed, firsthand, the productivity of their garden. However, don't be discouraged by this. How much more damage can you do to what's already been done? It is better to build your gardens and grow your own food, and make mistakes, than it is to do nothing and continue to support the promotion of vast monocultures.

Designers who have ambitions to produce designs for other people, whether they are paying clients or friends or long-suffering relatives, must have the practical experience in their own backyard. Nothing beats first-hand knowledge and acquired skills from gardening, designing and building your own systems.

Again, permaculture is not just about designing gardens. It is about designing human settlements. It is a plan that endeavours to maximise and enhance human interaction with the environment that surrounds them. This plan considers all facets of human existence. Coming to the realisation that changes are needed to the ways humans live, and then facing the bold step of acknowledging that we should do something about it, is crucial for our own survival on this planet.

Many people find it difficult to accept these ideas and change their outlook. But to embrace permaculture you have to change because it requires you to look at your life and lifestyle from a different perspective. Graham Bell, well-known permaculture teacher and author from the Scottish Borders district, told me that "permaculture is not a destination, it is a direction". You don't suddenly "arrive" at some point to finally declare "I'm a permaculturist". It is a life-long journey of change and growth.

We are only limited by our imagination. Each of us has the ability to make some difference in the world in which we live. There are no problems, only solutions, and sometimes we need to look at some things from a different direction, ignoring what we already know and stepping outside of our comfort zone. As they say, one person's rubbish is another person's resource.

Permaculture is also about designing stability and permanence. We try to use plants, on the least amount of land, that give the greatest production. While yields in a permaculture system can be high, much higher than natural bush or forest areas, there is a limit, no matter how well we design and how ingenious we are. Plants and animals have limits to their growth and production.

Having a diverse garden, meaning lots of different plants and animals in it, is no guarantee that the garden will be highly productive. Gardens may be only productive, in terms of high yields of carbohydrates, proteins and biomass, if they are consciously maintained. Left to themselves, neglected gardens quickly fall into disrepair. They may still be productive, in an ecological sense, but they may not necessarily be useful for the owner.

There is also a difference between yield and production. Garden plants can be very productive - they grow rapidly, produce lots of biomass as new leaves and extensive branches and root systems - but they might not produce a high yield. Yield, to me, is what you can actually obtain, by way of food primarily, but also by any other useable products such as dyes, medicinal treatments from herbs, or timber and firewood. While a garden may look like a jungle, you may not actually get much out of it.

Total yields in any system have both tangible and intangible components. We've briefly dealt with the tangible and those yields we can measure, such as mass of lemons produced by a lemon tree.

But what about those "yields" that we simply cannot measure. How do you measure a person's health and well-being and the pleasure they derive from their garden? These intangible yields have worth which we often seem to neglect.

How can you measure the effectiveness of your passive solar house, other than comparing it to the "average" consumption of a typical house elsewhere, or the energy bills you paid in a house you once lived in elsewhere? How well would a conventional house fare if you had built this type on the property right where your solar house is? These are system yields that defy measurement.

I used to think that we had to design the most intensive, productive system possible, but nowadays I think we should aim for a system that produces just enough - just enough of what we need and a little extra which enables us the opportunity for sale, exchange or to give away.

Many people think of food gardens and self-sufficiency when they think about permaculture, but permaculture is really about sustainable living. What you learn in the garden is closely linked with natural cycles, and it also provides the opportunity to take responsibility for growing your own healthy food. Furthermore, we mustn't forget that many plants in our permaculture systems may not be directly useful to us, but may be essential in the life cycle of insects, birds and other animals.

I don't believe permaculture should be seen as a way in which people can become self-sufficient. The emphasis should be on people becoming self-reliant, with positive interaction and co-operation between all members within the community.

It is true that the goal of permaculture design is to create an edible landscape. Like a natural forest, our gardens should contain a mixture of both different species and different ages. How to achieve this is known as succession, and it occurs in all ecosystems.

Ecosystems

Permaculture has been described as cultivated ecology, where humans deliberately develop cultivated ecosystems which are designed to maintain genetic biodiversity and minimise energy and matter inputs. An ecosystem describes an area and all that it contains - the living things and their non-living surroundings, such as the air, water and soil particles. For example, an examination of a pond or forest ecosystem would show the many food chains and other relationships between the organisms which live there.

In nature, most ecosystems are exceptionally fragile, and small changes, most often caused by human thoughtlessness, can destroy the system forever. This is due to the interconnectedness of the web of life. All things depend on each other in either direct or indirect ways. For example, some animals eat others, plants require nutrient wastes from animals to grow, and plants make the oxygen which all living things require.

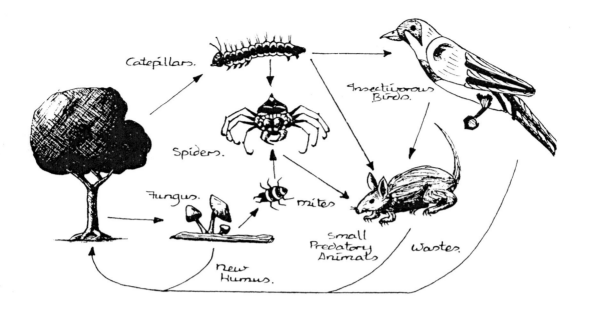

Figure 1.3 A forest ecosystem showing some of the feeding relationships between organisms and how these organisms interact with their environment.

Energy is continuously lost from all natural ecosystems and thus must be replaced to maintain the larger numbers of interactions that occur. The sun is the source of all energy on the earth and permaculture systems are designed to capture as much of this free energy as possible. Plants are nature's energy absorbers and storers.

Ecosystems also cycle matter. Stable, natural ecosystems have little matter input and output, meaning that organic material is not brought into the area or exported to other areas. Plants grow, die and thus return their organic matter to the soil. Permaculture systems endeavour to mimic this natural cycle by the deliberate intensive planting of a wide range of herbs, trees and shrubs. Animals complete the system so that a complex food web develops within the ecosystem.

We try to develop ecosystems which are self-perpetuating, and although they are dynamic, there is a balance of numbers of organisms within it. We say that the ecosystem is in equilibrium, even though small changes are always occurring.

Permaculture gardens change by ecological succession, but unlike nature, humans manage to direct the changes. Succession is normally the slow, progressive change in an ecosystem. When land is cleared the first plants are usually broad-leaved weeds (pioneer species) which cover, protect and shade the soil. Next come herbs and shrubs, and finally trees. Permaculture systems speed the succession process with planted herbs and trees so that the climax (final) community is achieved at a faster rate. The succession of plants and animals slowly changes and modifies the environment.

The early colonisers or pioneers, such as broad-leaved weeds, have special characteristics to survive inhospitable environments. These include:
- usually short-lived (either annuals or perennials only lasting a few years).
- not very competitive.
- unpalatable (unlikely to be eaten).
- prolific seed producers (or have persistent runners).

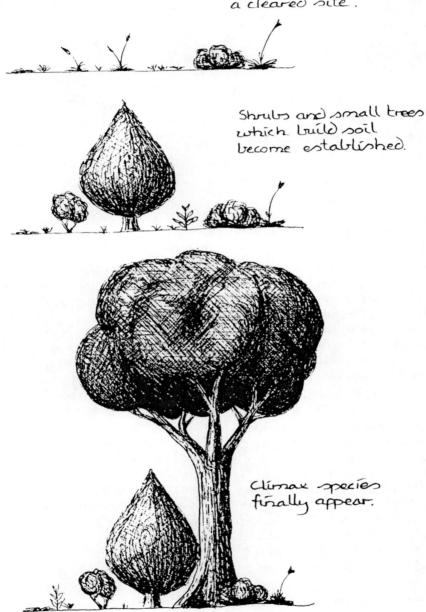

Figure 1.4 Succession is the gradual change in the numbers and types of plants and animals in an area over a period of time.

- defence mechanisms such as thorns or poisonous substances.
- tendency to be dynamic accumulators and to mine particular minerals from the soil.

Their role is to protect, cover and build the soil. The pioneers change the soil so that other species can follow and colonise the area too. As succession proceeds, the larger shrubs and trees, which are longer lasting and more competitive for resources such as light, water and nutrients, replace the pioneers, as shown in the figure opposite. Succession is discussed in further detail in Chapter 3.

I believe that we should do our designs to the best of our knowledge and ability and let nature take over. We might grow a nitrogen-fixing plant, such as vetch or clover, underneath a fruit tree, thinking that these will connect and all will benefit.

This naive view is what most of us hold. Nature is much more complex than this, and the relationships between plants, and between plants and animals, are very complicated; so much so that little is generally understood about how it all works. When we companion plant our fruit trees, we probably don't realise the many other connections and interactions that occur between these types of plants and others nearby, as well as their interactions with soil organisms. For example, some nitrogen-fixing plants make the soil non-wettable. Soil particles become coated with organic matter, which repels water. When you water your garden, or when it rains, water doesn't effectively soak into the ground. Water availability to your trees is thus reduced.

Sustainable land use

Sustainable land use occurs when the productive capacity of the land and its resource components of soil, air, water and biota are preserved, and hopefully enhanced.

The aim of your design is to develop ideas of sustainable land use, to the best of your ability, making the property as intensively productive as possible, minimising energy input or loss to the system, and hopefully allowing you or the property owner to work towards becoming partially self-reliant. The principles of sustainable farming include:

- self-sustaining systems must be developed. We want the land to be productive, year after year, but we have to look after it for this to occur.
- diversity. The property should be about one-third arable or horticultural and two-thirds stock and fodder. This is true for much of Europe and less true in Australia where a larger proportion for stock is usually allocated as most soils are poorer and rainfall is less frequent. These figures will vary, depending on the climate, soil and other factors, but they include a large proportion of trees (30%) which should be planted in every site.

The ratio of land use activities does depend on factors such as these, and examining the annual rainfall and soil type in a particular location will allow the determination of appropriate ratios of stock and other farming pursuits.

- net yield must be high. A good surplus provides income and this contributes to the standard of living.
- small size. This gives more productivity per unit area and is easier to manage.

One person can possibly manage 10 ha (25 acres) of intense farming, but would have difficulties looking after 40 ha (100 acres) with the same level of care.

- economic viability. Workers or

farmers must generate enough income to survive, and maintain and develop the property.
- maximum processing of farm products at the source. For example, milk can be made into cheese and yoghurt on the farm.
- both aesthetically and ethically acceptable. Consideration is given to the impact of buildings and lifestyle on the environment. However, functions generally come before aesthetics in any design system, and it is probably true to suggest that aesthetics are a by-product of good function and good design.
- conservation of wildlife, habitats and forestry. Some natural, untouched or restored areas should be set aside for wildlife.
- using appropriate technology - not necessarily heavy machinery. Simple, low energy-using or energy-producing technology is best.

Sustainability has an ecological component and an economic component. For example, when considering tree planting and revegetation work, you need to include the public benefit and environmental cost in the assessment, not just the cost-benefit ratio for the farmer or landowner.

Economic viability is ultimately dependent on ecological viability and both need consideration in sustainable land use strategies.

We must give an account of our ecological stewardship, and thus the guidelines for any ecologically-based design for sustainability would include:
- providing long-lasting future benefits, but meeting the immediate and current needs of people.
- protection and conservation of existing natural bush or woodland.
- strategies to improve soil fertility.
- maintaining or enhancing water quality, and harvesting and storing water.

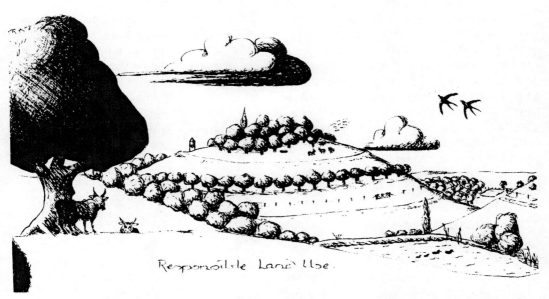

Figure 1.5 Responsible land use, while still reaping financial and environmental benefits, is the benchmark for sustainability.

- using appropriate species in the development. Local (indigenous) species are the first choice, but proven exotics, which have minimal impact on natural ecosystems, are acceptable.

  We tend to ignore, or maybe we are just not aware, that native forests can provide high yields of bee and animal forage, edible fruits and nuts, and good timber and firewood.

- using locally available materials and resources for buildings, fences and other structures.
- using appropriate technology for energy and power production, water movement and general property development.
- developing systems which require low maintenance and input, and which are easily monitored.
- consideration of the cultural, social, economic and legal aspects of the people.

For the systems we design to be sustainable they have to produce an equal or greater yield than that which is consumed or used in the system.

System yield can be defined or measured in terms of energy rather than biomass, which is limited. System yield is the surplus energy that is stored, or made available, in the system.

The total system yield is not infinite. Permaculturists are interested in obtaining a sustainable yield - a yield that still allows all things in the system to maintain themselves and survive.

The system becomes dynamic when there are a large number of interconnections between components, or there is a large amount of interconnectedness within the entire system.

There is also a large range of benefits - not cost related - that we always seem to ignore, yet should consider. For example, revegetation projects have:

- recreational value, where people can visit and use a reclaimed and restored area.
- aesthetic value, where landcare plantings can both look pleasant and serve many other functions.
- bequest value, where future generations of our children will be able to enjoy the area as well.
- intrinsic value, where it is recognised that all living things have some worth and have a right to exist. This is a powerful reason why remnant vegetation should be protected.

Sustainable farming practices and the concept of whole farm plans are further discussed in Chapter 11.

My notes

Things I need to find out

2 Maximising the edge

Land-based food production

Nature is full of patterns and shapes. You can find intricate spirals in sunflowers and galaxies, hexagons in bee hives and circles in rain drops and winds. Mathematical shapes like squares, prisms and rectangles are rare in nature. They may occur in some chemical crystal formations, but spirals and circles are much more common in the Universe.

Patterns are found in every known ecosystem, and the interface or boundary between two ecosystems is extremely rich in life - in complexity, abundance and variety. The interface between different surfaces, objects or ecosystems is known as the edge. It is where the air meets the water, the forest meets the grassland, the land meets the ocean, as well as the area above and below the frost line on a hillside and the zone around a plant root in the soil.

The edge is rich in life because it contains species typical of each side, such as those in a forest or grassland, and species unique to the boundary itself. Edges are very productive systems as there is more light available in these areas, a large variety of plants are found there, animals seek protection, food and shelter and make their homes there and the resources of two ecosystems are shared.

This is why permaculturists create designs with large amounts of edge and particular patterns; hence circles, keyholes and curved garden beds.

For example, as wind rises or is deflected upwards at the edge of a forest, it usually drops its load of dust, sand, leaves, water or seed.

Thus, the edge of a forest or heavily treed area is rich in nutrients and organic matter, and they are very productive.

Even so, there are some negative aspects to creating edge in gardens. For example, you can get weed invasion in newly planted beds in much the same way that you find in burnt and cleared bush or forest. The weeds (pioneers) quickly invade these areas.

Changing the shape of garden beds not only increases the amount of edge, and therefore production, it saves on watering. For example, circle garden beds are far easier and more economical to water with a sprinkler or spray system than a rectangular or square bed, as most sprinklers project a circular spray of water.

Convoluted garden beds as shown in the diagram below have a greater surface area or edge than the rectangular beds. More plants can be grown in these types of 'folded' garden beds. Common types of

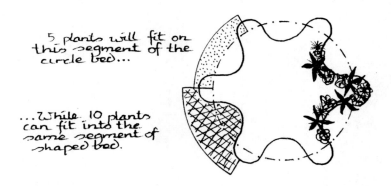

Figure 2.1 Changing the shape changes the amount of surface area or edge.

garden beds used in permaculture designs include:

(a) herb spiral. This garden bed is built like a pyramid, with a path or water line spiralling upwards. The circular base of the herb spiral is usually about two metres diameter and rises about one metre above the ground, but it can be larger and higher, or smaller than this.

A spiral is a useful pattern as the amount of edge and growing space increases as the garden bed climbs higher.

Figure 2.3 Circle garden beds are very common.

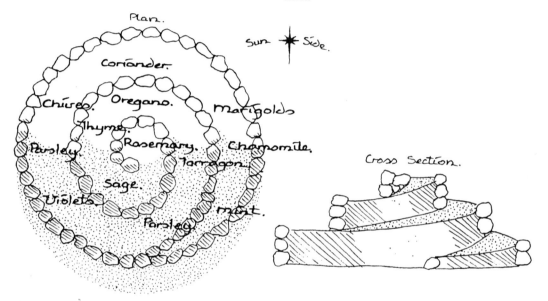

Figure 2.2 A herb spiral.

(b) circle. These are very common beds and they are usually between one to two metres diameter. If you make them too wide you cannot easily reach all plants, so these beds are deliberately small.

Some people make each circle garden bed a different theme. For example, planting tomatoes, basil and other Mediterranean herbs and vegetables as an Italian cooking theme, or having a circle bed for medicinal herbs, cooking herbs, teas and drinks (lemon grass, lemon verbena and lemon balm and so on) or pest repellent herbs. Herbs are good companions for vegetables.

Unfortunately, circle beds don't have a lot of edge, as circles have the least edge of any shape for their area of cover. We say that circles, and spheres generally, have the lowest surface area to volume ratio.

The other problem with a group of circle beds close together is that the space between them varies, making paths sometimes too narrow or too wide.

(c) keyhole. The "keyhole" can either be the path that you step into to reach your bed or the shape of the garden bed itself.

Keyhole paths allow easy-reach access to all plants.

Figure 2.4 Keyholes allow easy access to all parts of the garden bed.

(d) clipping and plucking beds. These types of beds contain vegetables that you can snip things from - maybe a leaf or flower - as you would for lettuce and nasturtiums. You can walk alongside plucking beds and harvest foods such as tomatoes, silverbeet, spinach, parsley, capsicums and many other useful foods. See Figure 2.5 which depicts a narrow plucking bed.

(e) narrow beds. Mainly designed and built for narrow garden places where you don't have much room, but what you have can be effectively used to grow vegetables such as onions and chives.

(f) broad beds. Many summer crops such as melons and pumpkin require room to spread. Once the initial mounds are seeded and the vine crops start to grow, allow them to fill the area, as this helps reduce weed infestation.

(g) vertical. Small-sized gardens may have to utilise vertical growing space. Think about using fences or trellis for vines (grape, pumpkin, cucumber, squash), berries (soft fruit, such as boysenberries and raspberries) and peas and beans. Trellises can be used to provide shade and create suntraps. Trellises range from wooden lattice structures to wire strung up between poles to allow a vine or creeper to grow up and into a tree.

(h) raised garden beds. Not everyone is able to bend over gardening all the time - and who wants to anyway? Building a garden bed off the ground is a simple solution. You can build gardens on top of

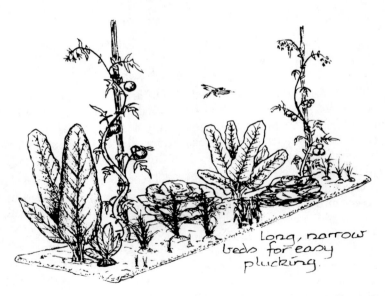

Figure 2.5 Narrow beds are ideal in small spaces. You can harvest from a plucking bed as you walk along the path.

Figure 2.6 Broad beds are good for melons and vine crops.

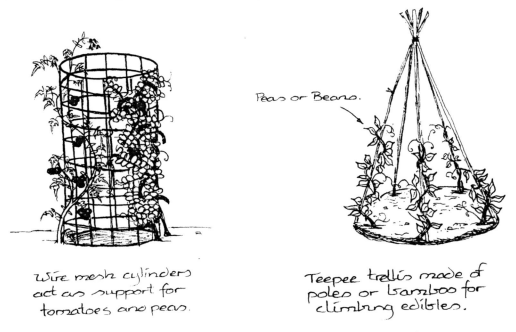

Figure 2.7 Use vertical growing space for greater food production.

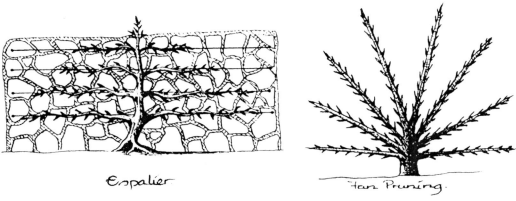

Figure 2.8 Espalier and fan pruning are techniques to utilise vertical and horizontal growth in fruit trees.

Figure 2.9 Raised garden beds can be built on any ground surface.

old, single, wire bed frames or box a garden with sleepers or concrete slabs. People in wheelchairs will be able to access and enjoy their gardens, and young children find it easier to work on these types of garden beds.

There is no limit to the diversity of plant and animal species that can be integrated in the terrestrial polyculture system, and the layout of the plants can vary also. For example, instead of planting in rows, with particular spacing, trees can be clumped one to one and a half metres apart, leaving available ground for interplanting.

Sometimes, orchard trees can be clumped. This technique is suitable for some nut and fruit trees, including coconut, edible palm species and banana. Planting pioneer leguminous trees, such as *Acacia* and *Albizia* spp., amongst the edible food crops further enhances the system. Underplanting your fruit trees with pest repellent herbs and other good companions makes these trees healthy and reduces the need for other types of pest control.

Water-based food production

Water is very important in permaculture. Water systems can achieve greater overall protein production per square metre of water surface than the same area in land systems.

Even small ponds can produce a steady supply of edible foods such as watercress, water chestnuts, taro and water spinach (kang kong).

Some of the most productive ecosystems in the world are found at the land-ocean interface or edge. These include mangrove swamps, which are on the land side of this interface, and coral reefs which are located on the ocean side.

Any aquaculture we develop needs to be set up like these ecosystems, with lots of edge, which not only contains plants and animals with commercial value, but a host of other organisms.

These are the basis of the many food chains that need to be developed and functioning if such a system is to work.

Unlike terrestrial polyculture, aquaculture polyculture should only involve a

Figure 2.10 Underplant orchard trees with herbs and ground covers.

Figure 2.11 In polyculture aquaculture many species are grown together.

limited number of species. While, in principle, permaculturists would like to make complex ecosystems involving literally tens of different species, it becomes increasingly difficult to monitor and maintain such a system.

Simple polycultures involving four to six species (at the most) is the ideal that you should deliberately introduce.

Thus you could have some crustaceans in the bottom of the pond, one or two species of fish as middle and top dwellers and several plant species that the animals feed on or shelter in, or both.

Many of the famous Chinese polyculture systems use herbivorous fish. The raising of carnivorous fish adds greater costs of production, mainly in feed.

The rest of the ecosystem develops naturally. The microscopic algae and bacteria, and the macroscopic insects, plankton and other invertebrates, all appear shortly after the pond is conditioned.

Conditioning a pond or dam involves adding organic plant and animal matter to make a rich detritus which is the basis of most of the food chains that become established in the aquatic environment. Small ponds, such as tyre ponds and bath-size tanks, are not conditioned in this way.

To condition a pond, a rough guide is as follows. Approximate amounts are given for a particular dam size and volume, so you can determine your specific requirements for your dam or pond.

For a dam size of 50 m^2 (e.g. 5 m x 10 m), volume = 100 m^3 (depth average 2 m), break up and spread two bales of straw and add two wheelbarrow loads of animal manure evenly over the pond.

Within a week the water changes to a darker colour and there is evidence of living things. The dam must then be aerated, continuously if possible, but at least for two hours, four times a day for three or four days.

The water must be moved, drawing up from near the bottom and spraying it across the surface. The water becomes aerated at the surface.

The water should start to turn a light green as algae become prolific. If mosqui-

Figure 2.12 The submersible pump inlet should point upwards so that it does not draw in detritus material from the bottom.

toes are a problem, continue with aeration and water movement. If the water doesn't become green, you may have to add a little more manure or some soluble phosphate fertiliser - about a tablespoon should be enough.

Once the pond or dam is conditioned in this way, you can then add the plants and animals to the system - butnot all at once. Plants first, especially the forage ones for the animals.

You may have to wait six months or more for the plants to become established before you introduce fish and crustaceans. These animals will devour, pull out and decimate plants that are not well established.

Don't put lots of deciduous fruit and nut trees around the dam, as large amounts of organic matter falling into the waterway will quickly pollute the system and all living things will literally die overnight due to lack of oxygen.

When fruit or leaves fall into water, bacteria and fungi break them down. Large amounts of these organisms consume the small amounts of oxygen present in the water, especially at night when all living things respire and use oxygen. The pond life dies.

Oxygen levels are also dependent on water temperature. As temperatures increase the amount of dissolved oxygen decreases. Ponds and dams containing warm water may need regular aeration if plants and animals are present.

Figure 2.13 First condition a pond, then introduce plants and finally animals.

In small ponds, where the water temperature can easily change, the average water temperature is usually about three-quarters of the air temperature.

For example, if the air temperature is 30°C the water temperature might only be 24°C. This is important, as most freshwater aquatic animals will die if the water temperature rises to over 30°C for a day or two, sometimes less.

Edge in aquaculture is equally as important as it is for land-based polyculture.

Figure 2.16 Fingers of land surrounded by water can be economically and easily fenced to reduce the impact of unwanted predators.

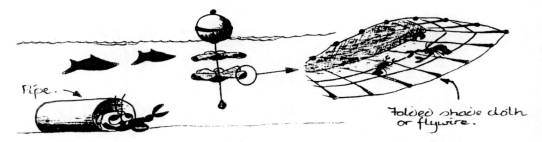

Figure 2.14 Hiding places are very important for territorial animals such as many freshwater crayfish species. Suspended, floating hides are used as protection sites when crayfish moult.

Figure 2.15 Change the shape of a dam to increase the edge and, therefore, the production of the system.

Ponds or dams don't have to be the typically-found square or rectangular shape. Adding curves and bends to the edges of waterways increases the surface area, and again, greater production and carrying capacity in the system is the result.

The chinampa system of Mexico and SE Asia, for example, is one of the most productive human-made aquaculture systems. Here, thin strips of water and land are interwoven to provide lots of edge for bird and other animal habitats, as well as for food production.

The edges of waterways are important development areas for many organisms. Fish, crustaceans, molluscs and amphibians all breed and grow in the shallow waters of waterways.

Reeds, rushes and sedges, and a host of other macrophytes, are found growing on or near the water's edge. These plants provide food for both aquatic and land ecosystems as well as shelter and breeding space and material for birds and other organisms.

The concept of maximising the edge in design is further illustrated by the shapes of windbreaks, houses and other structures which are discussed later in other chapters.

My notes

Things I need to find out

3 General design principles

Elements

We use the term "element" to describe anything we place in the design. This can be a plant, an animal, or some building or other structure. However, elements also include those natural features that may be present on the site. For example, numerous large rocks, clay and gravel (for mudbricks and rammed earth walls) and a patch of remnant bushland or woodland all impact on the final design strategies.

We use elements to provide the many functions we desire, and that make the property productive and beneficial to all things which are present. These functions include windbreaks, food production, fire control, energy production, and soil building and conditioning.

Each of these functions is served by several elements; in this way each element contributes to the whole system. If one element fails there are others to perform the function.

It is the large number of useful relationships which exist between the elements of a permaculture system that makes it so dynamic and functional.

Diversity in a system increases the likelihood of success. Our observations of nature's rich, biodiverse ecology confirms this.

It is because of this philosophy that we view organisms from a different perspective. We view weeds, for example, as pioneer plants, or we say things like "we don't have weeds, we have dynamic accumulators". In holding these views, we recognise and acknowledge the basic life ethic that all things are worthwhile and have intrinsic value themselves.

Knowing what elements we should place in our system is one of the hardest problems we have to solve. Designers basically need to know where things are placed and why.

To help us work out what we should develop or grow, we perform a Site or Element Needs Analysis.

Here, we consider all aspects of the element, such as its Intrinsic characteristics, Needs, Products, Behaviours and Functions.

As an example, consider the earthworm, which is the best animal to have in every permaculture system. A brief analysis is shown in the Table 3.1 below.

Integrating all elements into the design is the next step. We make sure that some elements can supply the needs of the earthworm and that the earthworm products are used in the system as well.

We try to use the products of one element to satisfy the needs of another element. For example, organic matter, as food for the earthworm, may be supplied by falling leaves, kitchen scraps or chicken manure and used paper.

The earthworm castings and vermicompost are also used in our design as they

Table 3.1 A needs analysis of an earthworm.

Analysis	Element characteristics
Needs	organic matter, oxygen, coolness, other worms and water
Intrinsic characteristics	breed or type (for example, red wriggler or tigerworm), size and colour
Behaviour	fast breeding, prefers cool, dark environment
Products	castings (worm poo), vermicompost
Functions	aerator of soil, breaks down organic matter, soil tiller

supply nutrients to our growing plants.

It is this holistic approach to design that makes permaculture unique. We try to develop an integrated, functional ecosystem that has a diverse number of elements (principally plants and animals).

Furthermore, judicious use of plants and other elements in our designs is also meant to cause less stress for animals. Placing other elements, such as fodder tree species and water near to your milking goat, for example, causes less stress in the daily life of the goat.

Zones and succession

Permaculture is based on conserving energy - not only the type you have to buy to run the refrigerator or television, but that of human energy.

The design is such that you minimise the energy and effort needed to walk in different directions to feed the chickens, milk the goat and collect the vegetables for the evening meal. These elements, along with others in the system, are placed relative to each other and relative to the number of times you have to visit them on a daily, weekly or some other basis.

Not all elements need attention all the time. For example, long-term (slow growing) trees need very few visits.

The positioning of elements in the design is known as zoning. Imagine five concentric rings around your house, called zones 1, 2, 3, 4 and 5, with zone 1 closest to the house and zone 5 generally furthest away.

We place different elements in each, depending on how often we have to visit them. For example, vegetables and herbs are grown close to the house in zone 1, while orchard trees are further away in zones 2 or 3.

The more attention an element must have, the closer to the house it should be. Generally, zones are not bounded by fences or delineated in other ways. Zones are simply a convenient way to deal with the

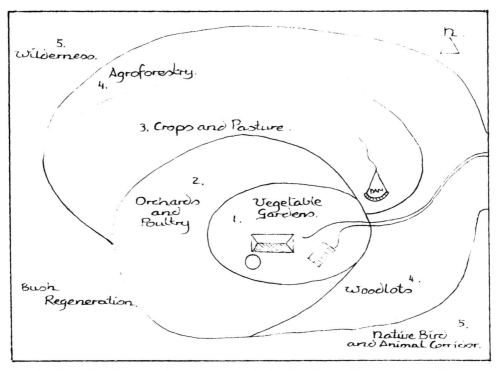

Figure 3.1 Zonation. Zones are imaginary boundaries that depict different parts of the property.

distance from the core of the system - you in your house.

Having a food crop or a herb garden close to the house invites good management, whereas distance encourages neglect. Things that are further away from the house, which you don't see or visit often, are often neglected. The emphasis here is to produce what you consume or to consume less of what you can't produce.

A brief mention should be made of the house area itself. This area is often called zone 0. The house needs to be integrated with the garden and the garden, in turn, can have a great influence on climate control of the house. This is discussed in greater detail in Chapter 8.

Each zone changes in the number and types of species, structures and other elements and strategies for implementation and maintenance. However, zoning is more a conceptual process where areas are allocated, but may change over time.

There is no magical shape for a zone. Figure 3.1 illustrates only one example. The location of zone boundaries depends on the size of the property and particular features. For example, zone 1, nearest the house, may be extended along paths or the driveway.

Generally, zone 1 is an intensive plant garden. Zone 2 also contains plants, but animals such as chickens or ducks are now introduced. A summary of the types of things (elements) typically found in each zone is listed in Table 3.2.

Suburban properties, which are usually 1000 m^2 or less, normally only contain zones 1 and 2. There is not enough room for elements typically found in the other zones, namely, orchards, goats, geese and other elements such as these. However, you may decide to dedicate a small area for zone 5 to encourage native birds and animals to visit the garden.

Small rural properties can support these additional elements, so you will often find zones 1, 2, 3 and 5. Broadacre rural is large enough to accommodate all zones, with the largest possibly being zone 4.

Now, putting this together, imagine this scenario. You leave the back door, collect the vegetables for tea or remove fallen fruit and dead leaves. Put these into the compost heap or take them to the chickens. Collect eggs, rake out any manure and uneaten food remains and place them in the earthworm farm (or compost heap),

Table 3.2 The kinds of elements in each zone.

Zone	Key features
0	House or other human living areas.
1	Intensive sheet-mulched food gardens, pond, shadehouse, greenhouse, rainwater tank, tool shed. Some fruit trees, such as lemon. Low windbreak around the garden.
2	Garden beds. Animals such as chickens or other poultry, earthworm farm, rabbits or guinea pigs. Aquaculture tanks or ponds. Hedges and trellising utilised for edge effects. Compost heap. Small orchard of fruit and nut trees.
3	Larger-scale orchards and geese, living mulches, goat pen, bee hives, fodder plants, windbreaks for house, firebreaks.
4	Woodlots (long term development), dams, agroforestry (extensive tree culture), shelterbelts, windmills, farm stock. Swales, drains, dams and other water harvesting strategies.
5	Wilderness, natural forest or bush. Catchment area and flora and fauna preservation. Wildlife corridors. Forest regrowth. Reforestation.

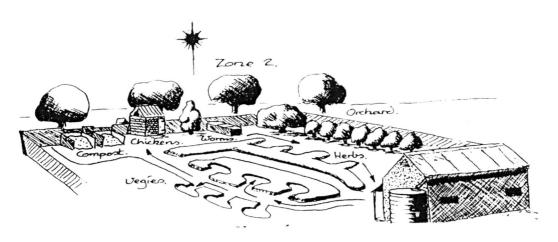

Figure 3.2 Elements are placed so that human energy is saved.

continue back to the house, collecting plucking greens or herbs such as lemon balm or lemon verbena to make yourself a nice cup of herbal tea. We design our systems like this to save energy.

Left to themselves these zones slowly change. In particular, areas which we designate zone 5 wilderness areas should be fenced off and left alone on rural properties. If we plant natural species back into these areas we are, in effect, planting an artificial forest, not a natural one. Provided that stock, weeds and excessive fire are excluded from these areas, natural bush and woodlands will develop themselves over a number of years.

The succession of plants in these areas follows a similar path, no matter where on the planet you live. For example, in the UK, pioneer species such as bracken fern, blackberry, gorse and broom invade the site. Soon, larger trees such as rowan, birch and oak start to grow amongst the pioneers.

These trees, in turn, shade out the pioneers and allow other small shrubs to develop. Some of these pioneers, especially the nitrogen-fixing ones such as gorse and broom, are short-lived, and either fall over or just die - releasing the stored nitrogen to the soil.

Each pioneer has its own niche or role to play in the ecosystem. Bracken invades land which is low in potassium or has been burnt. It "mines" or brings up potassium from deep within the soil.

Blackberry and gorse are very dense, prickly shrubs which prevent stock, deer or rabbits from attacking the oak, birch or ash trees growing up in the middle of the thicket.

In cool temperate regions, the soft berry fruits are the pioneer species. It is very common in the UK, for example, to see

Figure 3.3 Prickly brambles such as blackberry protect young trees from damage by animals.

blackberries, wild raspberries and gooseberries on the edges of woodlands.

Brambles can prevent the formation of low branches in climax species. The trees grow rapidly through the low-lying bramble bushes and then develop the characteristic round crown shape so typical of many trees.

Brambles also shade out grass from beneath larger trees. Many grasses release chemicals which inhibit tree growth and development and they should be used sparingly in most design systems.

In Australia, a similar pattern emerges. Fast growing, short-lived pioneer species of acacia and casuarina or allocasuarina, which are all nitrogen-fixing, establish in cleared bushland. These condition the soil and contribute to the growth and protection of the large climax species of eucalypts and banksias.

In the USA, nitrogen-fixing pioneers such as black locust, silverberry and mesquite, give way to the climax species such as walnuts, oaks, pines (e.g. redwood) and pecans.

Pioneers generally do not need mulching or much looking after - many are not fussy about the conditions they grow in. They prepare the soil for the plants that follow. Hopefully, once a design is implemented, the only thing you need to do is the fine tuning so that it progresses unhindered.

Sectors

Sector planning is another design strategy that is used when a site is analysed. Sectors consider the energies that move through a system. These energies, such as those of wind and rain, are directed, harnessed or used in the system.

Some of the elements in a sector are natural, such as winds, rain or sunshine, while others are human-influenced and artificial, such as aesthetic views and screens from neighbours.

To direct these energies or to counter their effects, we design windbreaks and suntraps, plant screen trees, and place deciduous and evergreen trees in particular areas of the property.

To know what kinds of strategies or elements we need to place in a design, we must have information about the property. Observation of your site is the most crucial part of the design process. You should make note of the sun angles, wind direction, areas of partial shade, sunny aspects, and damp and low-lying areas. The role of observation in the design process is discussed in the next chapter.

As an example of sector planning, consider sun angles. You may remember that the highest position of the sun in the sky (the zenith) changes from season to season. It may only rise to a height of 35° in winter but 85° in summer. This affects the

Figure 3.4 Winter shading - large trees block winter sunlight to the food garden. Summer shadows - the garden is still in full sunlight at midday.

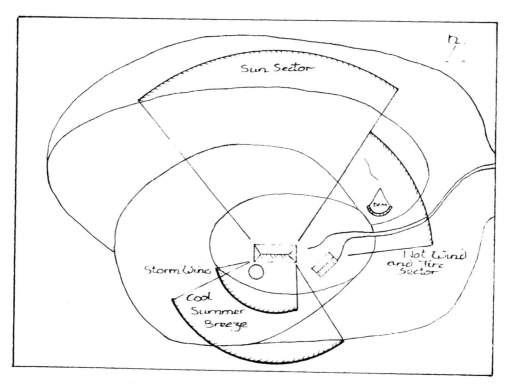

Figure 3.5 Sector plan for a typical property.

length of the shadow, which could be important if you want shade for the house or lots of sun for the food producing areas. Shadows will be longer during winter time, as shown in Figure 3.4.

Remember also, that the sun angle changes with latitude, so find out what angle the sun rises to in your location during summer and winter.

Sector planning often determines the position of windbreaks and garden beds, house location and the placement of many other elements in the system.

If we have a flood-prone area, we will plant trees that can tolerate inundation and heavy soil.

If we want to protect the house from winter storm winds and rain, we will design a windbreak to deflect and reduce the ferocity of the winds and adverse weather conditions. You can read more about the location and design of windbreaks in Chapter 11.

Microclimate

The term "microclimate" refers to the climate in a particular area which varies in its temperature range, humidity and wind intensity. For example, sheltered garden areas may have slightly cooler daytime temperatures and warm night-time temperatures, reduced wind velocity and higher humidity than other areas nearby or somewhere else on the property - and there may even be significant differences within the same garden.

You can easily create microclimates by building terraces, windbreaks and suntraps, or by planting particular evergreen or deciduous trees.

Creating warm or humid microclimates will enable you to avoid frost damage, prolong the growing season of your vegetables, cause the early ripening of fruit or allow you to grow other types of plants which may not survive in the normal climate. Even composting around the base of trees will generate heat and change the microclimate conditions.

With thoughtful design you can change an environment by the use of particular plants. Firstly, use plants native to an area or at least those which will survive in that particular climate and soil.

Secondly, by careful placing of windbreaks, shelter and irrigation, almost any type of plant can be grown in any area - provided, of course, that the plant's special requirements are met.

Some plants just don't grow in clay, yet others don't mind getting "wet feet" and they thrive in heavy soil.

Even so, many plants are susceptible to frost damage, and animals stress at high temperatures, such as above the mid-forties. Often it is more important to know the range of seasonal temperatures than the monthly average.

Ranges give an indication of the likely incidence of frost and the maximum temperatures that plants or animals may have to tolerate.

Similarly, it is probably more important to know when most of the rain falls and how much falls at this time than the yearly rainfall.

For example, if the yearly rainfall is 600 mm, 400 mm may fall in two months in winter and the other 200 mm mainly in late autumn or early spring, with little or none during summer.

This will determine how long plants, animals and humans will have to survive without water during this time, and hence there needs to be provision in the design to catch and store an appropriate amount of water to offset the drier times.

Frost

Microclimate design to counter the effects of frost is essential in many climatic regions. Frost is accentuated by land clearing and lack of vegetation.

Frost occurs when the land cools rapidly as it radiates heat and is most common in the very early morning, just before daybreak. Trees help prevent frost by trapping the heat as it is lost from the ground, thus preventing it from escaping altogether.

In this sense, trees help insulate the Earth. Many trees will also radiate infra-red heat at night from stored daytime sunlight.

Conversely, preventing heat from being radiated into the atmosphere also attracts frost. A mulch covering the soil prevents heat from being lost, and dew and frost settle on top of the surface.

Moisture needs to settle on a surface, usually just above ground level - either on leaves of a tree or mulch on the ground.

Figure 3.6 Heat loss from the ground is trapped by the tree canopy, thus preventing frost formation.

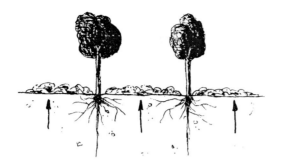 

Figure 3.7 Mulch keeps the heat in. Mulch prevents heat loss and moisture settles on the mulch layer. This will freeze if the temperature is cold enough.

Bare ground has less frost because heat is radiated skywards, and this is why it is best to have bare soil in the cold winter months if you want to keep some types of frost-sensitive plants alive.

However, there needs to be some consideration of the soil biota - those living creatures, such as earthworms, which will survive cold winters if they are kept warm by having a layer of mulch on the surface.

Cold air often settles on top of hills and in the valley floor. The cold air tends to fall from the hilltops to the valley and this can cause problems if obstacles trap this air. For example, inappropriately placed solid walls will trap cold air and frost as it moves downwards.

The secret in preventing frost is to move air so that water vapour can't condense and settle. This is why orchardists light fires in large drums to circulate air and why sprinklers are sometimes turned on. This unsettles the air and the water is comparatively warm, so frost does not form.

Knowing about the occurrence and regularity of frost is crucial to determining the length of the growing season. Late frosts often kill and set back newly-sprouted vegetables, so protection from frost is essential.

Some crop plants, such as oats and barley, are frost-resistant. However, others, such as rice and cotton, are frost-sensitive and soon die when temperatures drop.

Frost does have its advantages as well. Many fruit trees, such as stone fruit, require a certain number of chill hours (below 7°C) in order for them to complete their life cycle and produce flowers and

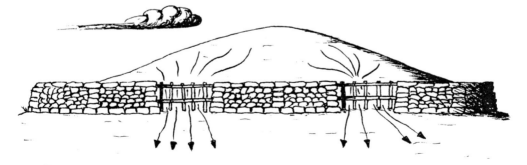

Figure 3.8 Walls should have gaps to allow cold air to continue to fall downwards.

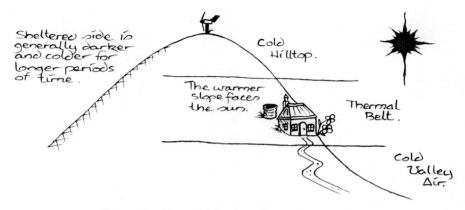

Figure 3.9 Frost forms on hill tops and in valleys.

fruit. Frost also kills many pests, breaks up compacted ground and provides large volumes of water to the soil and soil organisms when it melts.

It is not difficult to see why the best location for a house site is within the thermal belt of a slope. This is roughly the middle part of a slope.

Below this area, cold air settles in the lower valley areas and frost may be common. Observations during cold spells will allow you to determine the frost line, below which frost is common.

In ideal circumstances, wherever possible, buildings and house garden areas should be higher than this line, as little or no damage by frost can be expected. On the higher parts of a slope, cold air can also settle, as shown in the following diagram.

Aspect

Aspect, which is the direction that a slope faces, also creates microclimates. For example, in the southern hemisphere, south-facing slopes are generally colder and potentially have more shadow. North-facing slopes are usually sunny and warmer. Earth mounds surrounding a house, at

Figure 3.10 A hill with a sun-facing aspect has both shaded and sunny sides. Earth mounds (berms) are much smaller versions than this hill and are built around, or partly around, the house.

least on the prevailing wind sides, provide protection and different aspects such as sun-facing and shade-facing slopes for those plants that require these kinds of growing conditions. An earth mound can also be used as a windbreak and, coupled with trees, will protect the house from cold storm winds but may permit cool summer breezes to be directed toward the house.

Furthermore, the external temperature falls about one degree Celsius for every one hundred metres rise in altitude, so you would expect properties high above sea level to experience lower winter temperatures than coastal regions.

This doesn't mean that you can't use land which is high above sea level. There are a number of strategies to make the land warmer and to create warmer microclimates than the surrounding countryside.

In dry climates, where water is scarce, it may be better to design for coolness, rather than trying to make things warmer.

Reflection and radiation

Leaf colour affects the amount of light which is absorbed or reflected. We use this concept and plant shrubs and trees which have light coloured or silvery leaves, and waxy and shiny leaves, to reflect light and heat to create a warm microclimate. Leaves are specially adapted for different climates. For example, hairy, thick leaves are found in arid, sand dune and windy areas, while wax covered leaves are found in cold, high altitude areas. Use these features in designs for these types of climates.

We can also use structures, such as shed walls and roof areas, in much the same way as plants, which have light or silver-coloured foliage, to reflect sunlight into dark garden areas. Water surfaces can also be used to reflect sunlight and create a warm microclimate nearby.

Water plays an important role in creating and changing microclimates. Water has a

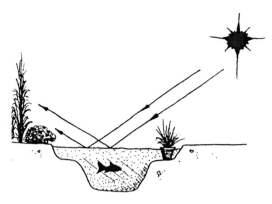

Figure 3.11 Water can be used to reflect sunlight and heat during winter.

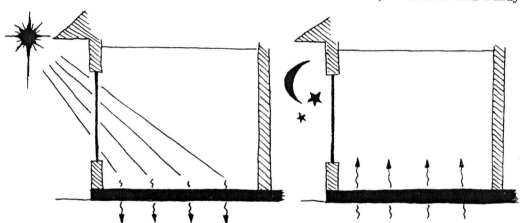

Figure 3.12 Floor radiating heat at night. Dark floors can absorb heat.

high thermal capacity, and dams and ponds can actually be used to store heat. Not only does water reflect heat and light, it slowly gains and then radiates heat (at a slower rate than land) and thus is warmer than nearby land at night. Gardens near waterways will be warmer too. As temperatures drop, heat is radiated from the water to the surrounding air and garden areas. This is why it is beneficial to plant trees nearer to coastal areas to take advantage of the temperature-moderating influence of a large body of water.

There is also higher humidity near the water's edge, which is good for subtropical trees. (I wonder if this is the reason why coconuts and palm trees grow on the water's edge?) Hot dry winds blown across water are cooled - before they hit the house! Conversely, cold winds can be warmed as well.

For the same types of reasons, we use dark-coloured bricks or slate in our house to absorb winter sunlight. These dark surfaces radiate heat into the house during night-time, making the house warm and comfortable. Again, grow frost-tender plants against or near heat-radiating earth mounds and solid walls.

Designing for catastrophe
Wind

You must consider the extremes in climate when designing. For example, little or no rain in summer, floods in winter and frosts in spring.

Sometimes, wind is the over-riding design criterion, such as in cyclone-prone areas.

Wind can be both friend and foe. Many plants will not get fungal disease if they are planted where wind often passes through. Some fruit trees, such as apples, pears and stone fruit, as well as grape vines, will be covered in mildew and other fungal diseases if they are planted in protected, shady and damp areas. You can solve this problem by shifting the trees and vines to a sunny area that has occasional winds passing by.

Wind is important for trees. Wind-stress strengthens the trunk fibres and over-protection may weaken the tree. Trees need to become hardy in their environment with little interference from humans.

On the other hand, strong wind can cause considerable damage to young seedlings and shrubs. Often, these types of plants need to be protected until they become established. Wind causes greater evaporation and transpiration rates, so that the plant and/or soil dries out quickly, finally causing the death of the plant. It is also important to plant windbreaks around dams to minimise water loss by evaporation. The evaporative cooling effect of winds is well known and temperature falls of several degrees are common.

It is important to note the direction of prevailing wind for design work. This is the direction of wind flow that is higher than the average. For example, if wind blows from the SE for 35% of the time (and the rest of the time less than this in all other directions), then this is the prevailing wind direction. The consequence of this is planning for windbreaks or strategies to harvest wind for energy generation and water pumping.

Fire

You have to consider the amount of useable, harvestable wind. In particular, wind associated with fire is a major concern. Don't underestimate the destructiveness and ferocity of fire. Fire spreads quickly through dry paddocks, along bush corridors and across the forest canopy.

Fire moves faster uphill than downhill, and even more so when fanned by strong winds. The most serious threat of danger is during hot, dry summers. For this reason, it is important to put as many wind and firebreak strategies on the downhill and windward sides of the house or sheds as possible.

Consider planting deciduous fruit trees,

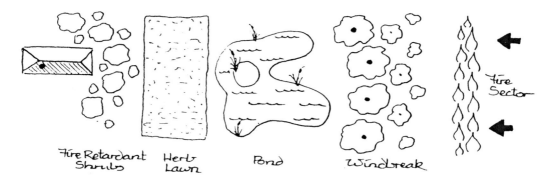

Figure 3.13 Some strategies for fire protection around a house site.

such as plums, apples and pears, and evergreens, such as carob and *Coprosma*, in fire-prone areas, as these types of trees are partially fire resistant and burn less easily than eucalypts, melaleucas and pines.

Other strategies include using grazing animals or the chicken pen to maintain bare earth, and constructing dams, ponds and roads.

It is a good idea, for example, to plant the deciduous stone fruit orchard in the poultry pen below the house as these trees do not burn well and poultry keep the ground bare as a fire break. Furthermore, dams or ponds are essential in this downslope area.

Since fire travels more rapidly uphill than downhill (twice the speed for each ten degrees rise in slope), then only fire-resistant trees and forests should be planted at the bottom of hills.

Generally, there is a greater risk and frequency of fire in grassland areas, followed by woodland and much less in wet forest. Designers need to be aware that in some parts of a country the risk of fire is much greater than elsewhere and more precautions need to be taken.

Furthermore, fires are spread by either burning along the ground or burning the tops (crowns) of trees. Crown fires are much more difficult to control and it is easy to see how houses can catch fire when tall, combustible trees overhang house roofs. Most fires in forests are spread via the canopy.

The intensity of a fire and its ease of spread depends on several factors. Some of these include:

- moisture content - dry timber burns much faster and at a higher temperature than damp wood.
- amount of fuel - intensity increases with increasing flammable ground litter.
- fuel size - small pieces of timber burn faster than large logs.
- wind speed - wind spreads fire at a greater rate than still air. Dry, hot winds off the land increase the fire danger more than cool, ocean or sea breezes.
- slope of land - as fire travels faster uphill there is a greater danger for houses on ridges.
- humidity of air - dry air enhances the burning rate.

The risk of fire can be minimised. There are many things we can do to reduce the fire risk. Some sensible strategies include:

- building roads as fire breaks. On small acreage properties, don't limit yourself to only one access road - the fire brigade may not be able to enter your property from this direction and your only escape is in the direction of the fire! This is poor design.

Figure 3.14 Windbreaks of fire retardant species deflect hot winds and capture ash and cinders. They produce a fire shadow to reduce the effects of radiation.

- siting dams and other water structures, such as rainwater tanks, on the fire risk side of the property.
- reducing the fuel load on the ground. Pine trees and eucalypts, for example, are litter accumulators and they build up a thick layer of dry litter underneath, which promotes ground fires. These types of trees should not be planted in fire-risk areas. Use low litter plants.
- planting green summer crops - grazed by stock or poultry. Grazing and browsing animals will also reduce the fuel load in every season, either by eating lots of the ground cover or by providing the manure, which allows micro-organisms to rapidly break down leaf and plant litter.
- windbreaks of fire-resistant trees and shrubs - to catch ash and burning cinders. Firebreak trees not only catch flying ash and burning leaves and twigs, they reduce the heat of radiation by producing a fire shadow.
  Radiation will often kill plants, animals and humans before the fire actually reaches them. Earth berms and buildings will also produce a fire shadow which will shade and shield organisms from heat radiation.
- poultry yard - poultry can keep an area bare and free of dry weeds and grass.
- placing greywater disposal area on the fire-risk side - this keeps the soil moist and vegetation green.

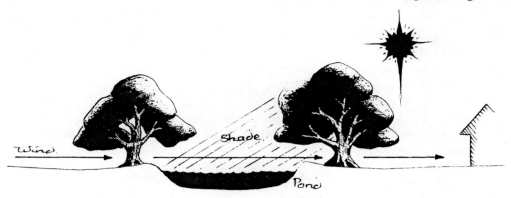

Winds can be cooled by passing them over water.

Figure 3.15 In hot climates, trees overhanging ponds will shade them and reduce evaporation. Winds can pass underneath the canopy and be cooled before they hit the house.

- planting a stone fruit or evergreen fruit orchard, with free-ranging geese to control grasses.
- sprinklers on roofs are essential in high fire-risk areas. Sprinklers, both on top of the house roof and in the surrounding gardens, must be put on before the fire approaches - very wet surfaces just won't burn and the fire cannot spread.

Fire does have its good points! We use it to cook and to warm ourselves. Lighting a fire in a 200 L drum (44 gal) at night when a damaging late frost might occur, helps to circulate the normally still air, preventing frost from forming and settling, thus protecting your frost-sensitive plants at blossom time.

Abundance of water

Water is the main limiting factor in our agricultural and horticultural endeavours, no matter where we live.

Too little and all living things die; too much and flooding occurs, which often has detrimental effects on plants and animals. Designers need to know details such as the highest known flood levels on the property to avoid potential problems in house siting, and the highest or estimated intensity of the precipitation rate to devise strategies for limiting run-off and erosion, and to allow maximum water storage without excessive loss as dams fill and overflow.

Other considerations

The elevation of the sun, the slope of the land and the movement of water and air all make a difference as to where we place elements.

The slope of the land also determines the kinds of plants you should grow or the activities you can do. For example, steep slopes should not be cultivated. They should be planted with trees and shrubs which will hold the soil together (what's remaining) and prevent further soil erosion.

Shade-facing slopes usually produce a greater number of chill hours which permits the growth of some types of fruit trees. Furthermore, trees in sun-shaded areas may be inhibited from flowering and leafing too soon, so there is less likelihood of damage from late frosts.

Slope also presents problems. As already mentioned, wind and fire both race uphill and wind over hills causes turbulence. Sun-facing slopes, while very important in cool climates, are more fire-prone than shaded slopes as the soil, plants and leaf litter dry quickly and, therefore, burn more easily.

Houses in tropical or hot areas obviously need protection from too much sun, and this will require strategies of shade trees nearby or directing cold breezes toward the house (wind funnels).

In summary, careful consideration is made for the placement of elements in our designs. For example, for effective plant establishment we need an adequate supply of water, protection from browsers, protection from drying wind and excess sun, good soil, and minimum soil erosion. Hence the need for water harvesting and supply measures, fences, tree windbreaks (hedgerows), and appropriate soil treatment and conditioning.

My notes

Things I need to find out

4 Steps in the design process

The design process

Bill Mollison, in his *Designers' Manual*, states that permaculture design "is a system of assembling conceptual, material and strategic components in a pattern which functions to benefit life in all of its forms."

Moreover, the human intellect needed in the design process distinguishes permaculture from conventional landscape practices and it is human imagination that is the key to permaculture design. We are only limited by our own imagination and life experience.

The best designers are those who have a wealth of practical experience. They know what works and what doesn't. They know which plant and animal guilds work well for that area and climate (see Chapter 6) and they are realistic in their expectations about implementing the design. Grandiose designs, which are totally inappropriate or expensive, are not in the best interests of all concerned.

The fundamentals of permaculture design have arisen from an understanding of the cycling of matter and energy in nature. Here, matter and energy are passed from organism to organism along various food chains. Organisms die and decay, and their wastes are also broken down, which releases stored nutrients to the soil. These nutrients can then be re-used by plants to continue the cycle of life and death.

Furthermore, permaculture designs are based on broad, universal principles which allow for local knowledge so that local species can be incorporated. This makes sense. Wherever possible, local resources and species, found in that soil type, climate and area, should be used. This is

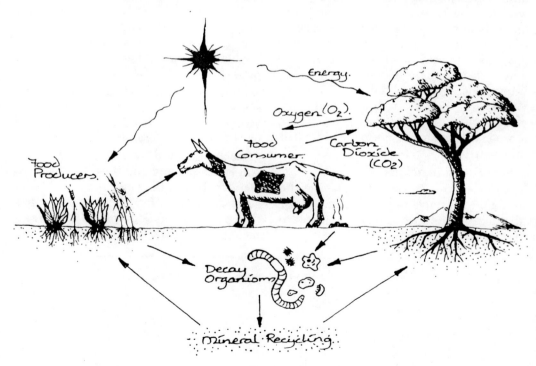

Figure 4.1 Matter and energy flow and cycle within an ecosystem.

both economically and ecologically responsible action.

In designing, we repackage or re-assemble components already existing on the property and incorporate new ones. Components are assembled and elements placed according to the function they perform. We use insight to develop unique and effective strategies. The design may examine many options and some decisions of particular options are taken so that they are definitely included.

Each permaculture design is tailor-made. It is the marrying and blending of what grows best in the particular area and soil with what the owner or gardener wants. Permaculture empowers people to solve their own design problems and apply solutions to their everyday situations. Designers have a responsibility to recognise which permaculture principles need to be applied to a specific problem or situation. Solutions may include the use of edge, patterns or guilds, and good designers see every difficulty as really an opportunity.

For any design to be useful it must have ownership by those living on the land. A design should involve everyone who lives there. A design is a collaborative effort, rather than the result of an "expert designer".

The trees are planted by the property owners and cared for by them to ultimately benefit themselves. The emphasis here is people. This is why participation in the planning and design is crucial to its success. Designing is a consultative process involving all of those living on the property and, sometimes, various advisers.

You can't successfully impose your ideas about a design on someone else's property. You have to work alongside the owners, so that they feel they have made a worthwhile contribution and they feel that they have ownership. In effect, we should teach people how to plan rather than plan for them.

Permaculture designing, then, is actively planning where elements are placed so that they serve at least three functions. We have already talked about how nothing is wasted in a permaculture system. The system we design also needs to use all of the outputs or products.

Pollution occurs when a system has excess outputs. The wastes of one element are used as the needs of another. The manure from the chickens helps the compost heap.

Dead leaves and fallen fruit keep the earthworm farm going, so that the castings produced will be the fertiliser or potting mix for your garden plants. If any product is not used, we have a potential pollutant and this is unacceptable in a permaculture system. Remember, we don't have liabilities, only assets.

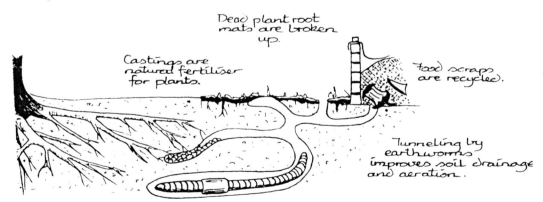

Figure 4.2 Elements serve at least three functions (earthworm).

A permaculture design is more than just a landscape plan. Maps, plans and overlays do not indicate or suggest the interconnectedness between things, nor can they deal with other aspects of permaculture such as the social and financial aspects of human settlements. However, a landscape-style plan does give some indication about dam and house placement, and the future location of swales and orchards.

You might think about the design as being a visual representation of the concept. Implicit in any design should be a number of energy harvesting and modifying strategies, a number of soil, water and land conservation strategies, a number of food producing strategies and a number of human settlement strategies, such as housing, shelter, village development and so on.

Designs always change and hopefully for the better. The design is the beginning point of the journey, and as new ideas and experiences develop, the design evolves as well.

Design considerations

While the client's wishes for a site are important, foremost is the consideration of the site itself.

What you are really doing is working out how the property will improve and become rehabilitated, because clients sell properties and move on. Look beyond the client and see the land.

When designing, all things that could affect the land should be considered. This includes sewage treatment, water storage, food production areas, placement of animals and the needs of humans. For example, you need to consider the characteristics of each tree you plant, such as spread or width, height, and whether evergreen or deciduous.

It may seem silly to plant a carob fifteen metres away from all other trees, but if you've seen a mature carob, you'll understand that they are often much wider than their vertical height.

There are some things on a property that you can easily change when you are working on a design and there are some things that you can't change. For example, you have very little control over the climate in a particular region, whereas you can change roads, location of dams and trees, and construction of fences.

Fences, by way of illustration, can be placed on contours and also along regions of soil change. Fences could then mark where sand areas change into clay areas and so on.

However, keep in mind that this may not be the best solution for a particular property. For example, keyline cultivation, discussed in Chapter 9, places soil type as a low planning priority, so fences may not need shifting.

Many things will affect the development and progression of a property. Some of these are beneficial and some are not. A list of these design considerations might include:

- erosion and salt scalds. Does water cascade over the surface or seep into the ground? What kinds of plants will grow in soil areas of high salt content?
- prevailing wind directions. Where will windbreaks be placed?
- views - those you want and those you want to hide. Where will you plant screen trees to give yourselves privacy?
- sources of noise and pollution, such as busy roads. Use trees to screen and to absorb pollutants from vehicles.

Permaculturists have to become accountable and ask themselves the questions "Do I create waste? Do I consume vast amounts of energy, electricity and heat in the home or vehicles?"

Figure 4.3 We have a responsibility to minimise our impact in the environment.

- recharge and discharge areas. Where does water enter the landscape and where does it re-surface?
- waterlogging and soaks. Are there areas that are permanently damp, clayey or waterlogged?
- slope of the land. Contours should be determined as water harvesting and water movement control are essential in dry climates.
- other components of the environment, such as aspect, amount of vegetation cover and soil type.

Our environment doesn't necessarily just mean the biological and non-biological components in the scientific sense, because, when we consider humans we also have to consider the economic, social and cultural aspects.

Furthermore, natural vegetation has intrinsic value and needs protection.

- amount of resource material available on site. Start by looking at the resources you already have or have access to. These vary from sources of mulch to clay and stone for building materials.

People argue that we ought to recycle more of the items we use daily, but I think that waste minimisation is a better, sounder path to tread.

It makes sense to use less than to recycle more. People often talk about "reduce, reuse and recycle". I would like to add "repair, refuse, re-think, re-assess, refrain, reject and reconsider".

Resources can be skills as well as materials. You need to establish what the client/s can do as part of the overall plan for the property.

You must also consider the needs of all occupants of a property, including children. Do children eat the same food as adults? Or is my family different from everyone else's? All need to be involved in the decision-making. However, making choices always involves consequences, compromises and trade-offs.

Design steps

The process of designing can be seen as a series of steps or phases. Like all planning exercises there is an order to follow when these things are undertaken. These steps are discussed in turn, but a brief overview would be:

1. Information phase - observing and collecting data.
2. Analysis phase - reflecting, examining and collating data. Recognising patterns.

3. Design phase - determining strategies, re-organising and placing elements in the system. Zoning and sector planning.
4. Management phase - implementing, priorities for implementing, and monitoring and maintaining.

Information phase

This is when you collect information about the site. Data can be collected through observation, or by examining maps and records, and by discussion with the property owners (and occasionally their neighbours).

It is essential to consider the needs and desires of the people living on the property and involve them as much as possible in the early stages of the design process. This can be as simple as getting them to list the resources available on the property or to write down their observations about storm and winter winds, flooding areas and so on. Good listening, to both land and people, is a hallmark of a good designer.

The most important information you need for a design is a map of the property. A scaled diagram showing the dimensions of each boundary and present location of the house, outbuildings and other structures, is used for the basis of the design.

Maps are very useful tools. You can obtain maps from local or state authorities which show the contours on a property (although these are not often accurate) and property shape and sizes.

The next requirement before the design actually begins is to mark and draw the existing landscape features - contours, ridges, valleys, streams and watercourses, rock outcrops and soil types, and including the location of the proposed house and outbuildings.

Contour lines on the map will determine where dams and water harvesting or movement will, or can, occur. Contours on the slope will also determine where retaining walls may need to be built and where particular garden beds or orchard areas should be placed. Using graph paper for your map is very useful and allows you to draw structures and ideas in proportion and to scale.

It is often useful to do nothing to your property for a year or so. At least see what happens during each season, such as which weeds flourish (as an indication of soil conditions), where the frost line is in the

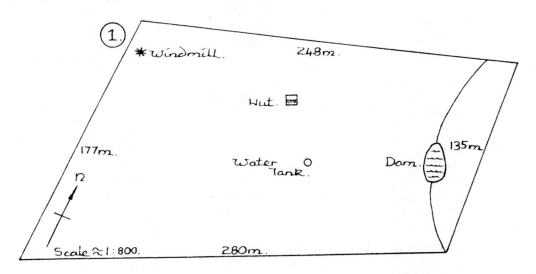

Figure 4.4 A map of the property is the first step in the design process.

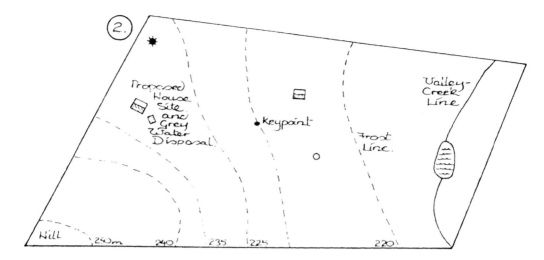

Figure 4.5 Develop the map by locating and drawing contours, landforms and other structures.

lower parts of the block and which are the prevailing wind directions.

Detailed and careful observation must be made of such things as wind directions, water movement on the property, amount (degree) of slope, sun angles during summer and winter, moist and wet areas, shady and sunny areas, changes in soil type, and natural patterns in any prominent land formations.

Information about rainfall, insolation days and seasonal wind direction and speed, can be obtained from sources such as the Bureau of Meteorology and, in some cases, the local post office or local government authority.

You should establish if you have the client's permission to spend money, on their behalf, for aerial photographs or cadastral and topographical maps, resource booklets of soils and plants of the area and so on.

Analysis phase

The analysis phase involves selecting and analysing the proposed elements in the system. A needs analysis may have to be performed on particular elements so that you can make decisions on which elements are best suited in the design.

Here we start to also consider zonation and sector planning. We draw and note the sectors for problem areas, such as

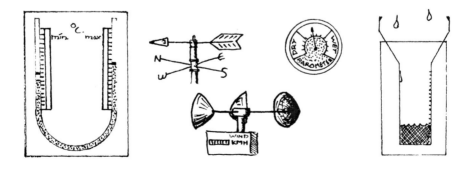

Figure 4.6 Collecting data and information about the site is a prerequisite to design.

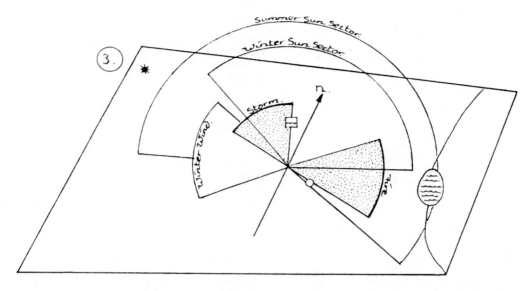

Figure 4.7 Draw the zoning and sector planning considerations on your map.

permanent waterlogging, occasional flood-prone areas and salt scalds. We also identify and mark storm and prevailing wind directions, fire danger directions and sun angles for the property.

Design phase

Most often a design is a map or plan which shows the placement of elements or components. This plan is usually a two-dimensional drawing, although a three-dimensional model, while taking a longer time to build and more resources to make, has many advantages. For example, a scale 3D model will allow you to investigate the effect of plant placement on such things as shadows and wind movements.

When drawing the design you can either use a series of overlays or several copies of the basic block plan showing dimensions, contours, natural rock formations, dams and other things which are fixed and mostly cannot be changed.

Some overlays can be used to show future developments, such as the location of new dams, fencelines, additional buildings, general plantings, windbreaks, and commercial enterprises.

Practical aids to help with design work include drawing concepts and garden areas on graph paper, using clay, plasticine, and wooden blocks to place structures in the design, and using a sand pit to shape slope, swales and dams.

All of these aids help us better visualise the design strategies, as seeing a design in three dimensions (3D), as we have just mentioned, is much better than in 2D.

A permaculture design might contain a large amount of information and either show or allow you to:

- identify food production areas, including zone 1 and 2 plantings, orchards and woodlots.
- indicate proposed drainage lines, swales and dams.

  This is essential for the control of water movement and harvesting, and to prevent flooding.

- indicate proposed revegetation areas - of natural bush.

  Zone 5 is allocated to the conservation and restoration of indigenous plant species.

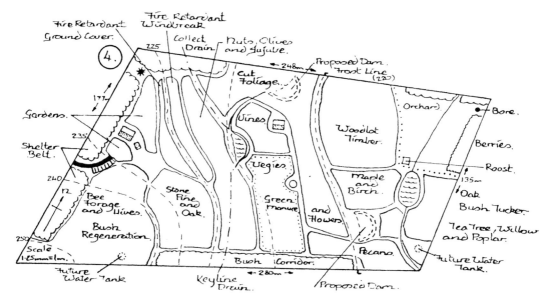

Figure 4.8 Developing a total design for Ian and Peter's property in Toodyay, WA.

- recognise the importance of sanctuary. People need to have some place they like to go and visit.

  This might be an aesthetic view or somewhere that has a particular calming feel about it. A design might include seats, a gazebo or log to sit on in this area.

- location and position of roads, paths and parking. Access roads should slowly wind upwards towards the house.

  Roads leading down to the house site generally cause heaps of problems, including water from rainfall and melting snow cascading toward the house.

  Having a driveway sloping away from the house will permit excess water to be diverted to garden and orchard areas (as well as allowing you to jump-start the car when the battery is flat).

- place sheds and house close to water and electricity supplies. This will minimise the costs of installation for these services.

The report, accompanying the design, will list the materials, components, strategies and priorities.

Design reports need ideas and hints about management, and land owners should be encouraged to undertake a permaculture design course so that they can understand the processes which occur and be able to make appropriate decisions about courses of action.

If you are producing a design for someone else, the information that you supply in your design and report **must** save the property owners **much** more than the fee you charge.

However, most designs for properties are done by the owners themselves. Few people are professional consultants and designers. A design report for a property typically contains the following information:

- design considerations. Write a brief appraisal of the kinds of things that need consideration in the design.

  This includes water availability and quality, the effect of native, feral

Figure 4.9 Access roads should rise toward the house.

and domestic animals on garden areas, seasonal climatic changes, landscape, client requirements for the property, orientation of the block and so on.

- site analysis - listing rainfall, temperature extremes and other aspects of the climate, soil type/s, existing vegetation, topography (for example, granite rock outcrops, slope of land and location of water courses) and general overview of the property.
- recommendations. A detailed list of changes to various parts of the property are given.

You might choose broad headings such as south side of block, north side, house site, orchard area, woodlot, zone 5, eastern hill slope and summer grazing area.

Explain and elaborate on these recommendations, including listing things that do not need to be changed on the property.

You need to explain why you are making certain recommendations. It may not be obvious to the client why you want certain plants grouped together, but if you say that this grouping will be mutually beneficial to each plant and there would be less risk of disease, then the client begins to understand the importance of design.

- priorities. You should list, in order of priority, the sequence to implement the design.

Identify the essential stages, such as water harvesting, which must occur early in the implementation process, and continue to list other important steps to the least important one.

- estimated costs for various stages of development.
- plant list. This contains details of suitable species appropriate for that climate and soil type. It should not list illegal or hard-to-get species.
- background information. Depending on client knowledge, you might include some information about

ecological and permacultural principles, such as guilds, stacking and succession.
- strategies for dealing with, or generating, income. Perhaps the client has little financial resources, but may have time and skills.

    An early, small income from activities on the site, such as seed collection and plant production, could be designed in.

    The client may be in full-time employment off the site and wants to reduce that. Can you plan to move them towards part-time work with an on-site income?

    Is there an opportunity for co-operative work in the local community? What does the client have to offer in exchange?
- maintenance of property. This includes information about the ongoing care needed for plants and animals.

    You may mention the particular nitrogen-fixing acacias (wattles) that you have included in the design (which will live for only eight to ten years and thus may have to be replaced); when they should add manure or mulch to growing plants; or how they can organically deal with fruit fly or other pests; how vegetables are to be replaced once they are harvested; and what types of husbandry goats or sheep need.
- resources. What resources does the local community offer? This includes people, who have particular expertise, such as dam building or knowledge of soil conditioning strategies, and nurseries where plants can be obtained.

    You can also list free sources of mulch, building materials or compost. Recommended reading and references can be included.

Management phase

Ongoing monitoring and review are essential, as unexpected impacts on soil fertility, plant and animal health and water quality can be noticed and design modifications made. It is important to develop a management plan which is reviewed and changed as the need arises.

There are many ways to monitor changes that occur on the property. This can involve periodically taking a series of photographs or transparencies (slides) which will depict growth and change in trees, and general development. Records of acidity (pH) and salinity levels in soils, dams, bores and other waterways can be kept and examined for trends. The data you collect from these types of activities will enable you to adjust the design strategies for these particular areas.

This phase also includes design implementation or execution which is covered next.

Implementing a design

Building gardens costs money. Too often we try to do too much too soon. We run out of energy, enthusiasm and money, and then time, to maintain the garden. Start off small, and when a particular area is set up, then move on and develop and build more gardens.

Work within your budget and plan to take a few years to implement your design. Start slowly from your house - work outwards from one zone to another.

Cost out each step or you may be disappointed and frustrated that you cannot complete the project because you have run out of funds. This idea of starting small and slowly progressing cannot be over-emphasised.

The implementation timescale should be based on economic reality - what you or the client can afford to do as time passes. Don't be too ambitious. Start small and meet with success. Then slowly expand as more resources, such as time, materials, money and energy, become available.

The order is: look after what we have first, restore what we can next and then finally introduce new elements into the system. When establishing a property the following are the sorts of things you need to do, not necessarily in the order of priority:

- water supply - earthworks, dams, swales, roads and drains. Priorities include the development of good water and appropriate earthworks such as drainage, dams and the foundations for the house.

  Earthworks are generally costly. There is a high cost for any sort of machinery and an operator, so plan to do as many jobs on one day as you can.

  For example, dig the power line to the shed and holes for a small pond or dam, level the ground for driveways, clear fallen trees, dig drainage lines to move water and so on.

- access roads and paths.
- structures - shelter. While machinery is available for earthworks do the house and shed pads.
- shelterbelts and windbreaks for gardens, orchards and planted areas.
- energy-producing or harvesting structures.
- plant procurement - nurseries, seed collecting and germination. You may need literally thousands of plants per hectare.

Implementing a design, even in a small backyard, can be daunting for some people. A lot of human-hours and human energy has to be expended to just build a few gardens.

This is where friends can be handy. Working with other people has many benefits and more can be accomplished while working as a team.

You sometimes hear of the term "synergy" when groups of people work together. This is where the sum of the whole is greater than the sum of the individual parts.

In other words, only a certain amount can be done by yourself, whereas working with someone else accomplishes more than two individual efforts. In essence, one plus one equals three.

There is a greater sense of satisfaction when you share the journey with others and the actions of many people can inspire you to continue to grow.

Remember, designers do not have to be experts on building houses and dams, identifying plants and animals, or driving heavy machinery.

A designer only has to examine the interrelationships between things and see possibilities to promote both biodiversity and productivity.

My notes

Things I need to find out

5 Basic tools for the designer

A designer's field tool kit

Every consultant needs to have some basic equipment that allows measurement of land size, acidity and salinity in water and soil, soil characteristics, slope and property orientation. Here is a brief summary of some of these types of items.

Tape measure

You need at least two tape measures. One, up to 10 m, for most urban backyard measurements, and one about 100 m for small acreage paddocks and some rural properties. I find a 30 m tape very useful. This allows you to measure urban property layout easily.

Figure 5.2 A longer tape measure is an essential toolkit item for larger properties.

Figure 5.1 Small tape measures are ieal for small, urban backyards.

You don't usually measure paddocks for rural designs. Farmers normally know the size of paddocks, or you can drive along the fenceline and record the odometer reading. Alternatively, use a dumpy level, which is described later, to calculate the distance. Rural designs are more concept-orientated and it doesn't matter if you are five metres out - you just plant an extra tree.

Piece of string

A 200 m roll of string, with knots every metre, can be used in open paddock situations. The string becomes tangled and caught on plants if you try to use it in someone's urban backyard.

The only other problem is that you have to mark the knots somehow, so that you know what the measurement is. The easiest way I've seen is to paint the knot and part of the surrounding string particular colours. For example, black for the first ten metres, red for the next ten metres, and so on.

Figure 5.3 A knotted piece of string can be used as a substitute tape measure.

Knowing the sequence of colours will enable you to calculate the distance. For example, it might be the blue section and knot number five along. This may translate to a distance of 85 m.

I prefer to simply pace out the dimensions of paddocks. If you practice your pacing steps along a long tape measure you can work out what distance you cover each step. I would recommend you practice a step of one metre. After a while, your paces become quite accurate, and certainly accurate enough for you to determine distances and draw a general map of a site area.

Penetrometer

A penetrometer allows you to gauge the depth of soil layers to parent rock, as well as some indication about soil compaction. You can make a penetrometer from a one metre length of 6 to 8 mm steel rod. Bend one end for a handle and sharpen a point on the other end, so that it easily penetrates the soil.

Generally, if you have difficulty pushing downwards it may indicate a hard pan clay layer below the surface or heavy soil. Rock is easy to determine - you push down and it just stops. Sometimes, you can hear the clunk as it hits rock.

Small, folding shovel or hand trowel

If you suspect heavy soil, hardpan problems or rock, you can simply dig a small hole and have a look. Furthermore, a hand trowel or small, portable shovel helps you to collect soil samples for analysis of pH, salinity, clay content and other soil characteristics.

Figure 5.5 Soil samples are often collected in the field and examined later.

Magnetic or directional compass

Every design should contain information about orientation. Normally, this includes a symbol marking the direction of north or south (wherever the main sun direction is, so that the property is sun-facing). Knowledge of property orientation allows easier understanding of wind and storm directions, as these are always expressed in reference to the compass points.

Figure 5.6 A compass indicates the direction and orientation of a property.

pH test kit or meter

There is a good range of kits and meters, which test the level of acidity or alkalinity in soils and water, on the market. Generally, you get what you pay for (where have you heard that before?), so buy a decent kit that you know will be accurate.

For the serious consultant or designer, it is worth the expense to buy an electronic

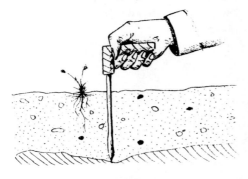

Figure 5.4 A penetrometer permits you to determine some subsurface conditions.

meter - especially since you can usually buy a variety of pH and salinity meters measuring a full range of levels and concentrations.

You will also need a number of sterile bottles or jars, with screw-top lids, to hold soil and water samples. Stick-on labels or marking pens can be used to mark the bottles with sample details.

Surveying the landscape

A variety of methods are used to measure slopes and levels. The simplest is the Bunyip or hose level, while more elaborate instruments include a theodolite and plane table.

Professional designers usually have a dumpy level as a minimum. While water levels, such as the Bunyip level, are very

Figure 5.7 Electronic meters are more accurate, but more expensive, than powder test kits.

Salinity meter

These hand-held meters are similar to the pH meter. However, most only measure a particular range of total dissolved salt (TDS) levels and you will, therefore, need to buy the one/s you require. Generally, a meter that measures within the range 0 to 20 mS/cm is suitable for most applications. Sea water is 35 mS/cm and most land properties have TDS readings in soil or water much lower than this. (Note: mS/cm means millisiemens per centimetre which is a measure of the electrical conductivity of a solution. 1 mS/cm = 550 ppm = 38.5 grains/gallon.)

accurate, their use in determining contours and slopes can be slow, especially over long distances or if a large number of points have to be measured. The following brief descriptions of instruments that measure distance and slope are listed in order of their increasing complexity and price.

Bunyip level

A clear plastic tubing or hose, or at least one metre of clear tubing at each end of a garden hose, enables levels to be determined. The hose is mostly filled with water, with enough space in both ends to allow for water fluctuation. Loose caps,

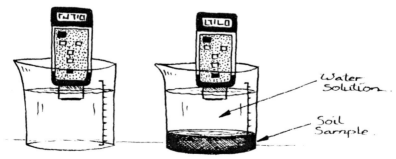

Figure 5.8 A salinity meter allows you to monitor both soil and water samples.

corks or hose clamps can be used to seal or pinch the ends of the hose so that water is held within the tubing while it is being moved around. Furthermore, using a hose with a bore diameter of about 12 mm (half an inch) permits faster equalisation of the water levels than a smaller internal diameter of say 10 mm (three-eights of an inch).

An "A" frame level

An "A" frame level is very handy for small areas such as suburban backyards.

It is made using two long pieces (2 m) of timber and one shorter piece for the crossbar (1 m).

A builder's level is strapped or tied to the crossbar. One leg is swivelled to a new

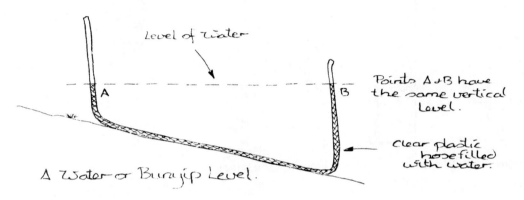

Figure 5.9 A simple Bunyip level made from plastic tubing.

Water seeks its own level. Hence, point A and point B, in the above diagram, are at the same vertical height. When applied to a slope, house sand pad or wall, the amount of fall can be determined over the distance between points A and B.

position such that the air bubble in the level remains in the middle. Both legs are then at the same vertical position.

An "A" frame can be pivoted around shrubs and other structures, so it has this advantage over conventional telescopic levels.

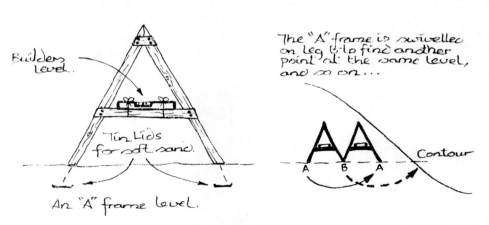

Figure 5.10 One leg is pivoted around the other. Points A and B are at the same vertical height when the bubble is central.

Dumpy level

Dumpy levels are used for greater accuracy. These are telescopic levels that are mounted level on a tripod. A staff is used for readings, and markings on the viewing lens permit not only measurements taken at hundreds of metres distance, but permit the actual distance to the staff to be calculated. You get both fall or rise and distance at the same time.

the top reading was 1.2 m and the bottom reading was 0.7 m, then the difference is 0.5 m. Multipling this by one hundred (this is universal for all such levels) gives a distance of 0.5 x 100 = 50 m. The staff is fifty metres away from the level.

More expensive equipment

Other more expensive surveying equipment, such as the plane table and theodolite, can be used if you want information

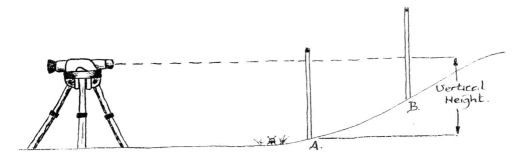

Figure 5.11 The difference between A and B gives the amount of rise in the slope.

The viewing lens usually has three crosslines. The middle one is the dead centre and this is used to read the staff height. Above and below the centre line are two shorter lines. By reading the staff measurements for these lines and calculating the difference between them, the horizontal distance to the observer holding the staff can be determined. For example, if

about slope, direction and distance. A simple plane table can be made using a timber board on top of a stool or stand. A large, 360° protractor is mounted on the board, so that measurements of angles to objects can be made.

This tool is ideal for quick, overview plans for a design, as it permits you to measure the distance and angle of any object to a

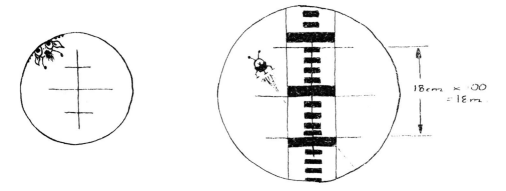

Figure 5.12 The viewing lens has three cross-lines. These help determine the distance to an object. Multiply the difference between the top and bottom lines by 100 to give the distance to the staff.

fixed point. By redrawing on a sheet of graph paper, the plan will show whatever objects are fixed for a design.

A theodolite is used by professional surveyors and is not discussed here. You would use a theodolite if you were surveying a large area for contour banks or for a village development.

Figure 5.14 A scale ruler is a must for designers.

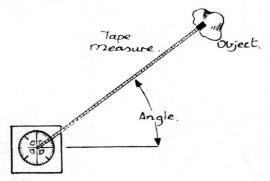

Figure 5.13 A simple plane table is a useful design aid.

A designer's drawing tool kit

Drawing a design requires some skills and some tools. These include a range of pens and pencils, other stationery items, and mathematical aids such as protractors and a drawing compass. You don't need an expensive drafting table or a drawing board, although both of these would be very handy. Some of the items for a standard toolkit are listed below.

Scale ruler

Drawings should always be done to scale. Many urban designs can be drawn on an A4 or A3 sheet using a scale such as 1:100. This scale means that one centimetre on the drawing represents 100 cm, or one metre, on the property.

Rural designs use a larger scale. For example, 1:1000 or more, depending on the actual size of the land you are drawing, is a common scale.

A scale ruler has various scales already imprinted on the ruler edges, and using one saves time and lengthy calculations of the sizes of objects and distances which have to be placed onto the drawing.

Light box

This is a box, normally wooden, which has a clear or translucent glass top. A fluorescent tube or incandescent light globe (40W) is housed under the glass. When it is switched on, the light shines through the glass so that copies of design drawings can be easily traced.

This box is also handy for viewing slides. Just tip the slides over the glass surface and you will be able to view many slides at once. This allows you to quickly sort and order the slides for a presentation.

Stationery items

Keep a good range of standard stationery items such as lead pencils, eraser, coloured pencils, sharpener and fine felt tip pens.

Pens and coloured pencils are used for shading and marking different regions of the design.

Make sure that you have a straight, unchipped ruler, a clean eraser, a new sharpener and a few felt tip pens for labelling and inking in the main design lines.

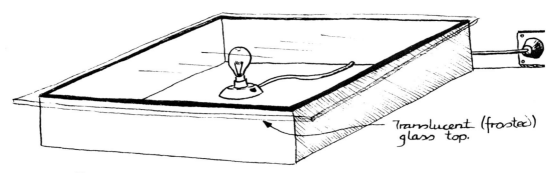

Figure 5.15 A light box allows you to make copies of basic maps and designs.

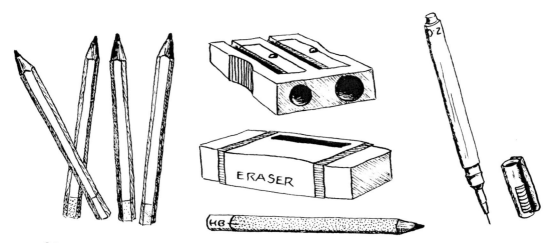

Coloured Pencils, Sharpener, Eraser, Graphite Pencils and Felt Tipped Pens.

Figure 5.16 Some of the stationery items you will need.

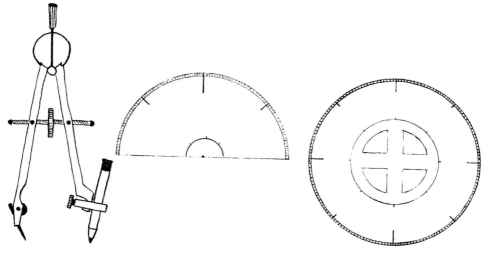

Figure 5.17 A protractor and compass are always used in design work.

Mathematical drawing aids

You will often need a few drawing aids. These include a protractor, Mathomat™ and drawing compass.

Protractors measure angles, and the skill of using a protractor and a compass is essential in design work. You have to be able to draw in slope (on elevation drawings), various shapes for buildings, rainwater tanks and other structures, and mark out sector angles.

Mathomat™, Mathmaster™ and Mathaid™ are common commercial mathematical drawing aids which contain a range of shapes, such as circles, squares and rectangles, as well as a protractor, curved surfaces and ruler. They are invaluable for design work.

Drawing paper

Designers need a reasonable range of graph paper, tracing paper and normal white bond art paper.

Sizes up to A2 (twice A3 size) are useful, but you can get by with just A3 and A4 paper sizes for most design work.

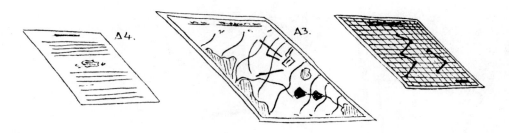

Figure 5.18 Make sure you possess a range of paper sizes for design work.

My notes

Things I need to find out

6 Basic principles of garden building and management

Introduction

This is not a chapter about how to build sheet-mulched garden beds. While it does have some hints about garden building techniques, the chapter focuses on design principles such as integrated pest management, stacking and guilds, as well as discussing the importance of soil.

We have already discussed how a permaculture design is a combination of techniques and strategies - how we build the gardens and why we place them there. Keeping this in mind, we should now realise that the garden areas should be:

- mainly perennial - mostly herbs and some vegetables, such as eggplant and globe artichoke, which last several years.
- self-perpetuating - allowing nature to take its own course by letting some vegetables self-seed each year. This may not be possible in very small gardens.
- diversified - place in the system as many different plants and animals as possible. Diversity is the key to successful gardening and forage systems.

The garden beds we build are organic ones. This means that we don't advocate or use chemical sprays for pests, we use compost and mulch as natural fertilisers for our growing plants, and we practise sound garden management and husbandry to minimise disease.

Organic and biodynamic farming systems have been shown to be viable, although they may not exceed or match the yields of traditional chemical farming methods. Organic produce usually gives a higher return for the crop, and with lower input costs, these food production techniques are seen as appropriate gardening and farming strategies. Organic and biodynamic food may not be produced in large quantities, but the quality is far better and more nourishing than chemically produced food. The secret to healthy food is healthy soil.

Building healthy soil

The nature of soil

A plant's ability to ward off disease is strongly dependent on the health of the soil. Soil is alive! Healthy soil contains a good balance of nutrients and elements, living organisms, humus (decaying matter), water, air and soil particles. This allows plants to obtain the necessary substances they need for growth, repair and disease immunity.

The mining of minerals from the soil by plants and the subsequent replacement and cycling of these minerals, with the creation of productive healthy soil, is one of the key design strategies in permaculture. The soil is the key to successfully growing food crops.

Soil is a combination of different particle sizes. It is a mixture of sand, silt and clay, the differences of which are shown in the following table.

The best soil for gardens is loam which contains about 20% clay, 40% silt and 40% sand. Loam also contains organic matter. This is essential for sustainable fertility, as it improves the structure, field capacity, exchange capacity and many other features of the soil.

Table 6.1 The classification of soil particles.

Particle	Size (mm)
clay	< 0.002
silt	> 0.002 < 0.02
sand	> 0.02 < 2.0

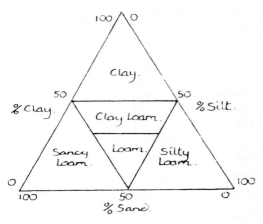

Figure 6.1 The composition of different soil types.

Clay soil is usually heavy, becomes waterlogged easily, and any holes dug for plants create wells full of water which eventually kills most trees. Clay is important in holding and then releasing water and nutrients to plants, but too much can create poor soil structure, depending on the chemical nature of the clay. For example, if the clay forms an impermeable barrier, it waterlogs easily. However, if the clay forms a crumbly structure the soil may be rich in nutrients.

It might be all right to say that you can easily change soil, but if you have lots of clay the expense of adding gypsum or sand or breaking the clay up can be prohibitive, depending on the local availability of gypsum or the area that needs treatment. Small areas where the house vegetable gardens are can be treated, but large areas would be impractical.

Clay has a high cation exchange capacity. Ions loosely held by the clay particles are exchanged with others from the surrounding soil and plant material. The exchange capacity of the clay depends on factors such as the soil pH, the types of ions held by the clay and the amount of organic matter present. For example, in highly acidic conditions (low pH) iron and aluminium ions are released by the clay, resulting in potentially toxic levels of these elements for plants in the soil. Figure 6.2 shows how zinc and calcium ions, for example, are exchanged with hydrogen ions from the plant roots.

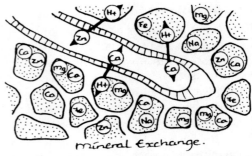

Figure 6.2 Clay binds positive ions (cations) strongly to its surfaces. These cations can be progressively released to plants by exchange with hydrogen ions.

Sand doesn't have these cation exchange properties anywhere near the same extent. It can be gutless, with low water-holding capacity and high leaching.

Sand has good drainage. The larger particles, and large pores between these particles, allow water and air to move through it easily.

Silt has a balance of the properties of sand and clay, with reasonable water-holding and water-releasing ability.

The other major component of soil is organic matter. Land owners often have to work to increase the organic matter content in the soil. Once the soil becomes "alive" again with the proliferation of organic matter and organisms, such as bacteria, fungi and earthworms, then nutrients become available to plants, plants are healthier and resist disease, and crop production increases.

Furthermore, more water can be stored in the soil. Even sandy soils store about 0.8 mm of water in 1 cm of soil.

However, soils with high organic matter can store 3.5 mm of water for every centimetre of soil, so there is a very good case for building up our soils with humus and organic matter.

Why mulch?

Mulching is one way to improve soil fertility which mimics natural forest systems. Organic mulches of chipped plant material, for example, can be spread about 5 cm thick over the garden beds. The mulch slowly breaks down and the action of micro-organisms and macro-invertebrates, such as earthworms, releases nutrients to the soil.

Mulch protects the soil, keeps the soil cool, allows greater water infiltration and reduces water loss by evaporation. Mulch also prevents erosion during thunderstorms with torrential rain, and the lower temperatures under the mulch keep microbes and earthworms happy. The soil should always be covered, and areas for growing living mulches, such as mustard and clover, should be considered.

Mulches can be made of both living and non-living material (organic and inorganic, respectively). Rocks and plastic can be used to cover the ground, but mulches of organic matter, including newspaper and cardboard, and shredded plant material, are much better in most cases. Organic mulches are best as they will, in turn, break down and provide greater levels of nutrients for plants. (Using rocks as a mulch is a design strategy in desert regions.)

The most important contribution of organic mulches is the lignin and other substances that they contain which are transformed into humic acid. It is the continuous production of humic acid that has sustained the fertility of rain forests for thousands of years due to its powerful chelating properties. That is, nutrients are held within the humic acid structure and thus are made more readily available to plants.

A large range of mulches is usually available. Each has its own particular use and you should use the mulch best suited for the job. For example, if you want to get rid of a lot of invasive weeds, then black plastic or some other lightproof membrane is ideal, while to cover an established garden bed, full of herbs and vegetables, then only a plant-based mulch, such as shredded tree prunings, should be used.

Wherever the ground is permanent sod or mulched the soil structure improves. However, where the surface is tilled or left bare by using herbicides, soil structure becomes unstable, with low porosity and higher risks of erosion and degradation.

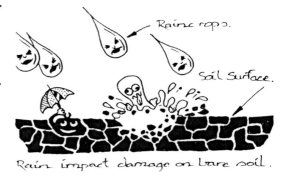

Figure 6.4 Water dropping onto bare ground causes soil to be lost by erosion. Mulch protects the soil and minimises topsoil loss.

Figure 6.3 Weed-free, deep litter leaf fall from *Acacia saligna*. Mulches reduce weed growth.

Mulch is great for almost every situation. However, a mulch layer on the garden beds in frost-prone areas during the colder months is not appropriate, as we have discussed in Chapter 3. Mulch prevents heat loss from the soil but frost tends to settle on the mulch and plants in the garden. Bare ground is a better alternative for these times because heat can radiate from the soil surface at night, minimising frost damage.

Soil conditioning and treatments

Permaculture is about restoring and rejuvenating the land, not mining it. Even though permaculture concerns itself with holistic, integrated design, it is not important to include every known strategy for improving the soil in the design, nor could every idea be put into practice.

However, sometimes soil needs to be changed. You may often notice that particular weeds are flourishing where you don't want them to be, or your plants look yellow and weak.

These weeds and many other plants can be used as indicators of soil health. Their presence suggests that the soil could be acid, wet, clay-based, sandy, nutrient deficient or some other condition. For example: in Western Australia, sorrel (*Rumex* spp.) and soursob (*Oxalis* spp.) suggest acidic soil; blady grass (*Imperata cylindrica*) or barley grass (*Hordeum leporinum*) suggest salt contaminated soil; while castor oil plants indicate recently cleared land and bracken fern a recently burnt area. The following table summarises some of the plants used as soil indicators.

Once an assessment of the soil has been made you can then decide how the soil needs to be conditioned or treated. Soil conditioning is the gentle changes you make to the composition, nature and structure of the soil. Frequently, it involves loosening the soil and/or adding amendments to change acidity, alkalinity or chemical composition. A full soil analysis of the types of minerals present is the only means of accurate soil nutrient amelioration and balance.

Soil conditioning can occur by several methods. For small urban properties, a garden fork pushed into the soil and moved slightly to and fro will cause air and water to freely enter the soil. Alternatively, deep-rooted plants, such as daikon radish, comfrey and dandelion, or larger tree spe-

Table 6.2 Plants as soil indicators.

Plant	Soil indication
Clover (*Trifolium* spp.)	low N
Salad burnet (*Poterium sanguisorba*)	alkaline soil
Fat hen (*Atriplex hastata*)	high fertility
Stinging nettle (*Urtica urens*)	cultivated soil, high fertility
Sheep sorrel (*Rumex acetosella*)	sand, acid soil, low in magnesium
Plantain (*Plantago* spp.)	cultivated, wet, clay soil
Wild radish (*Rapranus raphanistrum*)	low fertility
Dock (*Rumex* spp.)	wet, acid soil, low in magnesium
Horsetail (*Equisetum* spp.)	wet, clay, acid soil
Bracken fern (*Pteridium aquifolium*)	low K and P, acid soil
Chicory (*Cichorium intybus*)	cultivated soil, clay
Barley grass (*Hordeum leporinum*)	salty soil, water table near surface

cies, such as *Acacia* spp., will have the same effect, but slower.

Finally, let earthworms do the job for you. Place a layer of compost or mulch on the soil surface and let the earthworms do the rest. Earthworms are nature's gardeners.

In some countries, such as those in the UK, Europe and America, large animals such as moles and gophers, and insect larvae, supplement earthworm activity by their burrowing habits. In the drier areas of Australia (and in some other countries) where earthworms are seldom found, ants replace earthworms in the ecosystem. Ants fill an important niche in the cycling of matter.

For small acreage and broad scale farms, mechanical aeration with chisel ploughs (for example, Wallace plough and Yeoman's plough), or even a stump jump scarifier, can be used. These slice through the soil, cutting deeper each season, breaking up the compacted soil without turning it over like a conventional disc plough.

Mechanical machinery should not be used on steep slopes. Instead, the use of deep-rooted trees, such as pine and oak, will improve soil structure. These types of trees create soil while others, such as eucalypts, deplete the soil of its nutrients.

Ploughing should be done along the contours as part of the keyline system of land management, which is discussed in more detail in Chapter 9. If you wanted to plant a tree shelterbelt you would make only one run at the deeper cut (50 cm or more). Chisel ploughs can be used, but it is more common to "rip" the ground to a depth of about one metre using a bulldozer, pulling one to three rip tynes. This single, deep cut allows seedlings to become better established by sending down roots more easily in search of water.

As much as possible, minimum tillage should be practised. Heavy machinery and large numbers of stock easily compact some types of soils and this needs to be examined. Unfortunately, minimum tillage often implies the use of herbicides. By planting clovers and other leguminous plants, and using crop and soil management strategies, herbicides are not needed.

Many farmers (and gardeners) plant green manure crops which are easy to grow in most local conditions and which produce

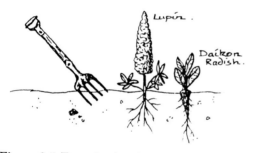

Figure 6.5 For suburban backyards, a garden fork or deep-rooted plants help improve aeration and drainage.

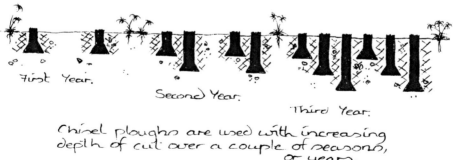

Figure 6.6 Chisel ploughs are used, with increasing depth of cut over a couple of seasons, on both small and large farms. This improves aeration and drainage, breaks up compacted soil and permits plant roots to penetrate deeper into the soil.

large amounts of plant biomass. These plants can be turned into the soil to improve its quality, and increase nitrogen and other nutrient levels. Other types of plants, called cover or catch crops, trap nutrient run-off and protect the soil. These plants can be harvested, by slashing for example, and used as a source of mulch and for compost making. Table 6.3 lists a few examples of green manure and cover crops which can be grown in particular seasons.

Finally, there are soil additives that you can use to ameliorate the soil. You might want to change acidic soil to become neutral, improve drainage and water and root penetration in clay, or provide particular natural fertilisers to improve soil fertility.

ing, chemical damage, over-burning, erosion and nutrient exhaustion, so will the human society, which over-works the soil, collapse.

Integrated pest management

Your garden should be a place of harmony between all of the elements in the system. Why is it that many gardens are riddled with pests and everything else has a feed of the vegetables you planted? You never get to pick fresh fruit and vegetables, so you just give up. Sound familiar? Maybe it is time to try integrated pest management.

Integrated pest management is a holistic approach to pest control. It is used in conjunction with the five principles for a

Table 6.3 Green manure and cover crops.

Season	Plants
Summer	cow pea*, lablab*, Japanese millet, sorghum, vetch*
Autumn	lupins*, canola, faba beans*, oats, field peas*, barley
Winter	oats, canola, mustard, faba beans*, barley, field peas*, rye-corn, sub-clovers*, medics*
Spring	cow pea*, vetch*, canola, sorghum, pinto's peanut*, mustard, buckwheat, lupins*

* legumes. All other green manure crops listed in the table are non-legumes.

For example, areas near coastal regions often have sands which are typically leached and low in nutrient content. Sometimes, these sands lie over limestone and are thus alkaline (high pH). Some amelioration, by adding acidifying substances such as sulphur and slow-release fertilisers to improve the nitrogen and phosphorus content, needs consideration before planting is undertaken. Some of the more common soil treatments are shown in Table 6.4. Many others, such as magnesite and rock phosphate, are not discussed.

So important is the soil to our very wellbeing, that I believe when the structure of our soils collapse because of over-stock-

healthy garden. These are:
1. Develop a sustainable polyculture.

The concept of sustainable polycultures was discussed in Chapters one and two. Basically, we want to grow a range of different plants that have different functions (niches) in the garden. For example, scented herbs and attractive flowers can be used as borders of garden paths. You could consider beds which have a seasonal theme such that different beds flower in spring or summer, or arrange plants such that there are always some in each bed which are flowering all-year-round. These herbs and plants may also repel pests and some will have medicinal, culinary or other

Table 6.4 Soil amendments.

Substance	Uses
Gypsum	improves drainage in clay and supplies essential protein-building "sulphate"
Dolomite	changes acid soil to alkaline, contains calcium and magnesium
Limestone	changes acid soil to alkaline, contains calcium
Rock dust	ground rock for sources of most nutrient elements
Blood and bone	organic fertiliser, high nitrogen and phosphorus content
Seaweed (kelps, not sea grass)	good source of trace elements, all-round fertiliser, high potassium
Worm casts	excellent, balanced fertiliser
Sulphur	changes basic (alkaline) soil to acid
Wood ash	from fires, contains high potassium levels (caution: very alkaline when fresh)
Lime	changes acid soil to alkaline, contains calcium
Sheep and pig manures	high levels of nitrogen, phosphorus and potassium
Cow and horse manures	slow release fertilisers, lower nitrogen content
Poultry and pigeon manures	very high nitrogen content. Pigeon manure contains the highest broad spectrum of all trace elements
Bentonite	expansive clay to increase water holding capacity of soil and cation exchange capacity of light sandy soils

uses. Flowers are also important for adult ichneumon wasps and hover flies which must have nectar before parasitising pests with their larvae. The orchard area is another place where polycultures can be developed. For example, you could underplant orchard species with:

(a) companion plants for pest control - such as nasturtiums, tansy and rue. Plants should complement others nearby.

(b) ground cover to smother weeds and utilise the area - such as sweet potato, melons and pumpkin.

(c) green manure crops to slash or turn in - such as clover and mustard.

There are many other ways in which complex ecosystems and polycultures could be set up. Why not experiment!

2. Use a diversity of strategies.

Some pests find food by the chemicals produced by the host plant. If you plant lots of the same type of vegetables together (as a monoculture) then the chemical signal is much stronger. This means you attract more pests.

Getting rid of garden pests naturally is difficult. You must use a range of strategies or you will be doomed to failure. The next few pages illustrate some ideas.

Organic growers use crop rotation as a strategy against pests and disease. The rotation of annual crops limits pest and

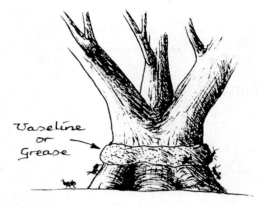
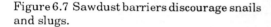

Figure 6.7 Sawdust barriers discourage snails and slugs.

Figure 6.8 Grease on the base of trees discourages climbing insects. It is best to place the vaseline on a band rather than directly on the bark.

disease attack and reduces over-exploitation of particular nutrients which would be mined year after year from the same soil. Many pests have part of their life cycle in the soil. By changing the location of particular plants each year you can disrupt this cycle. For example, in the same patch of ground you might plant carrots in the first year, then tomatoes in the second year, then silverbeet in year three, potatoes in year four and so on.

3. You don't have too many pests - you haven't got enough predators!

Aphids might be a problem for lemons or roses but they are seen as an opportunity

the garden and let nature control the pests.

4. Don't allow any ... icides near your garden.

Pesticides, insecticides, fungicides and weedicides kill useful predators as well as the pests. Avoid them! There are lots of alternatives.

For example, if mildew is a problem look for the cause. Consider shifting the plant into a breezeway or don't use sprays or sprinklers for watering. If you have to treat a plant for mildew use a "tea" of

Figure 6.9 Crop rotation breaks the life cycle of some garden pests.

by ladybirds. This philosophy ensures that we see the solution, not the problem. We look at problems differently and, in this way, decide that we don't have a snail problem, but we do have a duck deficiency.

Permaculturists design for predators. We develop strategies to attract predators to

some combination of chamomile, stinging nettle, comfrey or horsetail. All of these contain silica which inhibits the growth of fungus.

5. Nurture the soil.

Get the plant nutrition right and you eliminate most disease problems. It really is as simple as that. Healthy plants living

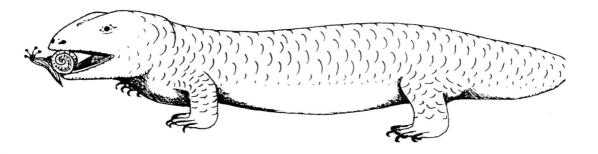

Figure 6.10 Use natural predators to eliminate pests.

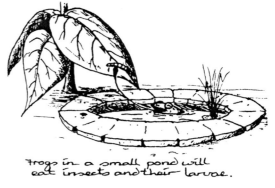

Figure 6.12 A pond also helps in pest control.

Figure 6.11 Umbelliferous plants attract predators to the garden.

in healthy soil do not get diseased easily and can ward off disease. There is much we do not know about how plants resist disease organisms, but we do know that they can, given the correct minerals and nutrients they require.

Too often we forget that in the living world, and in our own garden, we are the

Figure 6.13 Adequate pest control requires a range of different strategies.

Figure 6.14 Make pest traps in the garden. This is much better than spraying pesticides.

59

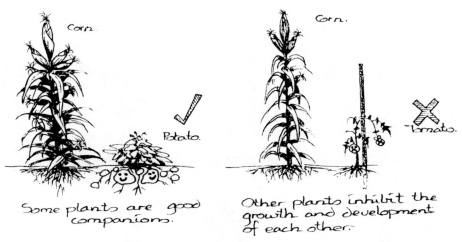

Figure 6.15 Companion planting ensures a healthy garden.

most serious pests. We compete for food with a whole host of other animals, and we must become smarter about how we can obtain and satisfy our needs and look after the other creatures at the same time.

Stacking

Many native peoples throughout the world practice sustainable forage systems. Vines, shrubs, herbs and trees are all grown together in a technique commonly known as stacking.

Stacking also occurs in nature. In any natural forest, you will find plants stacked together - tall trees have understorey shrubs and small trees, and grasses and herbs occupy the ground level. Below the ground, root and tuber plants proliferate.

Stacking allows as much as possible to grow in the smallest possible area. Dense plantings such as these suppress weed growth and soil erosion, and ensure that all ecological niches are filled by plants you need for your garden.

The degree to which stacking occurs depends on limiting factors such as the amount of water, light and nutrients. You would expect, for example, to find a denser grouping of plants in a well-watered garden than in a natural, dryland climate area.

Stacking uses vertical growing space more effectively. You should arrange or stagger plants according to height, tolerance to shade and so on.

For example, grape vines can be grown over a fig tree. Pumpkins and melons can be grown up and over a shed wall and roof - all utilising the free space available in a

Figure 6.16 Practice sound management techniques in the garden.

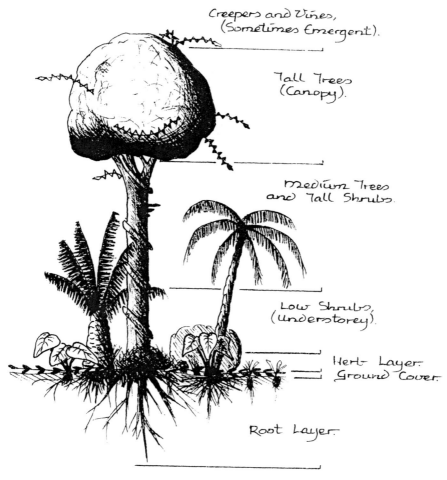

Figure 6.17 The principle of stacking.

system. You can use trellis, fence, tree trunks and walls to obtain more vertical growing areas.

Using the principle of stacking allows other possibilities. For example, you should aim to have successive crops of edible foods throughout the year - a variety of fruit trees such that you are always picking some fruit any time, any season. Furthermore, you don't have to wait to plant the next season's vegetables until you've cleared and finished picking that season's crop.

Time stacking is a technique where new seedlings are planted or come up as the previous crop is finishing.

There is less waiting time between one picking time and another. For example, you can grow silverbeet, onions and globe artichokes together. The silverbeet grows the fastest and is harvested within eight weeks. The artichokes take about four months to grow and flower, and finally the onions plod along and can be harvested after six months.

After the silverbeet is harvested, you can replant new seedlings and so the process continues. I will add that silverbeet can be harvested over a long time if you only pick one or two leaves from each plant at any one time. New leaves will regrow and so harvesting continues for quite some time.

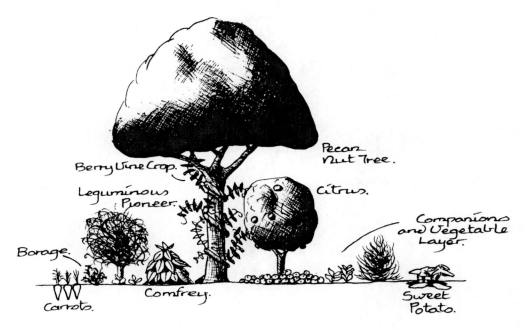

Figure 6.18 An example of stacking in a temperate-Mediterranean climate.

Guilds

Designing leads to the selective placement of objects, both living and non-living, in the system. The placement of living elements occurs as we try to maximise the benefits to each species. In this way, guilds are formed. A human-made guild is modelled on the functional, symbiotic diversity we so often see in nature.

Guilds are thought of as harmonious assemblies of several species around a central element (either a plant or animal). For example, for an orange tree, herbs such as lavender and rue would be planted as understorey to repel pests, and nasturtiums as ground cover (also repels insects) and to smother weeds and grasses. Clover and vetch can provide nitrogen in the soil, and an albizia tree is also planted nearby to attract ladybirds to eat aphids.

Ducks or chickens can be occasionally let in to eat slugs, snails and insect pests.

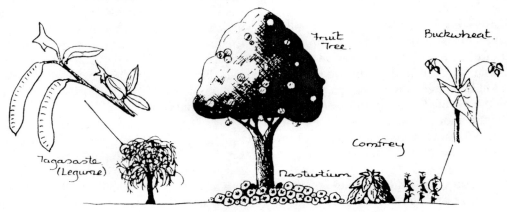

Figure 6.19 An example of an orchard guild.

(Poultry will devour nasturtiums and graze clover, but leave most of the other plants mentioned.) The albizia will also provide wind protection, nitrogen for the soil and mulch for the orange tree. This is an example of an orchard guild.

A guild, then, is an assembly of many different organisms, both plant and animals, which complement each other. It is an extension of companion planting which many people are familiar with, but it can involve animals as shown in the rock guild below. Here, a lizard eats pests in the garden, then seeks shelter in the rocks which, in turn, absorb heat and radiate it to create a warm microclimate for growing plants. This system is mutually beneficial to the plants and animals found there.

Figure 6.21 Shopping bags make ideal garden beds.

Figure 6.20 A rock guild.

Other tips for gardeners

Here are a few hints on particular aspects of gardening. Designers will be able to use these ideas as well as incorporating some of them in their own gardens.

- The Clayton's garden bed when you don't have a garden bed. Grow your vegetables in plastic or hessian shopping bags (double plastic) with holes in the bottom. This conserves more water than a conventional garden bed.
- There is a need to collect seed from your vegetables and herbs. In the garden, some of these plants should be allowed to complete their life cycle so that they flower and seeds are produced.
- Plant trees and understorey shrubs at the same time. In many cases understorey shrubs cannot establish themselves under existing tall trees.
- Garden beds should only be as wide as your reach, so that you don't have to walk all over the beds to harvest. Often, a bed one metre wide or one and a half metres wide will allow you to have access all the way round the bed.

Paths are also narrow, usually up to half a metre wide. Sometimes paths have to be wider for wheelbarrow or wheelchair access.

- Bare-rooted tree species are much cheaper and readily available. Many deciduous trees, which are dormant in winter and can be transported without soil, are sold bare-rooted. They are planted straight into the garden at the onset of winter.

As the winter breaks and the warmer spring weather arrives, these trees start to bud and grow.

- Bear in mind that you have more success with some vegetables than others. If this is the case, you may

have to change your diet and learn how to use more of these vegetables and less of the ones that you can't seem to grow.
- Sometimes, wherever mulch is used the soil becomes non-wetting. Water simply runs off its surface and doesn't penetrate.

To solve this problem you have to get water into the root zone. A piece of pipe or modified cool drink bottle can be positioned alongside the tree and drippers or other irrigation devices can be placed into the bottle. Water percolates down, deep into the soil, minimising loss by evaporation in dry areas.

Alternatively, water repellency is reduced by adding and mixing the clay bentonite to the organic mulches you place around the base of trees.

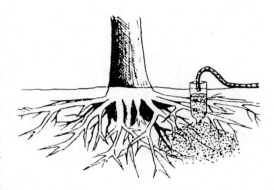

Figure 6.22 Plastic bottles can be recycled as water suppliers to the root zone.

My notes

Things I need to find out

7 It's all a matter of location

Location and climate

As you go from the equator towards either of the two poles, the number of indigenous (native) species decreases. The equatorial region has a rich tropical diversity, while the polar regions have much less variety. This phenomenon is due to the climatic changes that occur as you go from the warmer equatorial region to the much cooler polar regions.

Before we rush into discussing how we assess a property so that you can buy the ideal site for a permaculture farm, we need to examine some basic ideas about climate and location. Knowing about these concepts will enable us to make informed choices about various aspects of a potential permaculture site.

The local climate is largely determined by factors such as latitude, altitude, topography (such as hills, mountains, valleys and waterways), vegetation cover and closeness to the sea or ocean.

The daily and seasonal change in temperature, for example, is due to the angle of the sun's rays to the Earth's surface. As the latitude changes from the equator towards either pole the sun's rays become less steep. Angled sunlight is spread over a larger area, thus the temperature is low. When the sun is directly overhead the amount of light and heat is concentrated on a smaller area, which causes a warmer temperature.

The measure of the amount of sunlight reaching a particular point on the Earth's surface is called insolation. This is recorded as the number of watts/square metre, and at midday could be as high as 1000 W/m^2.

The value of the sun's insolation allows us to make predictions about the effectiveness and usefulness of solar equipment in different parts of the world, and about how much energy solar appliances can be expected to extract or harvest from incoming sunlight.

The angle of the sun also changes seasonally as the sun appears to move from one hemisphere to the other. The angle of the Earth's tilt on its axis and the direction of the Earth to the sun causes the seasons.

Here, the altitude of the sun (angle above the horizon) during winter may be only 35° but 80° in summer. At the same latitude, the azimuth angle, that between the north and the direction of the sun, may be 60° for mid-winter, for example, but 120° mid-summer. These seasonal changes are considered in permaculture design and were discussed in Chapter 3 when we dealt with sector planning.

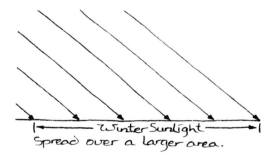 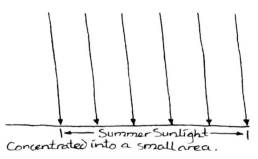

Figure 7.1 The temperature in a local area is partly due to the angle of the sun's rays to the Earth's surface.

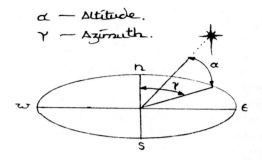

Figure 7.2 The difference between the azimuth and altitude of the sun's position.

Furthermore, greater condensation is possible as trees provide large surface areas for water vapour to condense. Forested mountain areas have permanent fog, mist or cloud on their peaks. In some places, condensation contributes more than rainfall in the total amount of water reaching the Earth.

We can think of a forest of trees as essentially a lake above the ground. Trees often contain more than 80% water and they continue to pump large volumes of water into the atmosphere. Not only do trees store large amounts of water in their trunks and leaves, plant roots also hold water in the soil, usually as a thin film around each root fibre.

Large trees can lose thousands of litres of water each day. Much of the rain and cloud formation comes from the trees in a forest. You can imagine what the effect of

What we call climate, the daily and seasonal changes in temperature, rainfall and so on, is influenced by the water cycle, and the importance of trees, in providing water to the atmosphere through transpiration and evaporation of water from plant surfaces for cloud formation, cannot be underestimated.

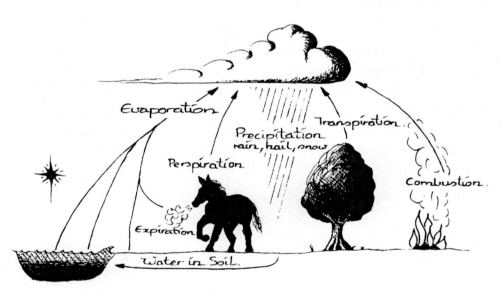

Figure 7.3 The water cycle.

It is imperative that no more forest areas are cleared and that literally billions of trees are replanted, especially on hill slopes, as these play an essential role in the water cycle. We have the ability to change and increase local precipitation patterns simply by planting trees.

forest clearing would have on climate patterns. Fertile land is slowly turning into desert.

Densely-planted trees also soften the force of rainfall and greatly reduce erosion and run-off. In fact, light showers may not even reach the ground and the water

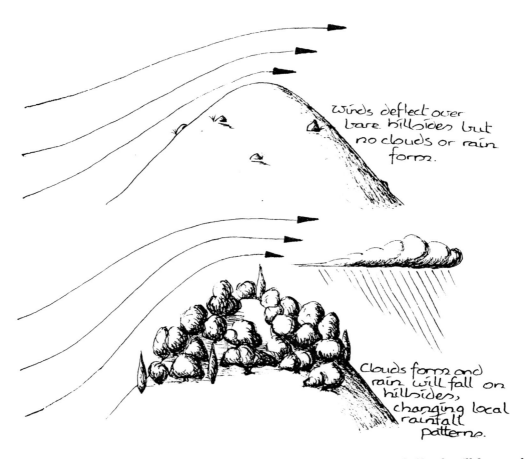

Figure 7.4 Wind deflects over bare hilltops but no cloud or rain is produced. Clouds will form and rain will fall on forested hillslopes, changing local rainfall patterns.

remains as a fine mist in the tree canopy, causing a humid atmosphere.

In cold climate (temperate) regions, such as the UK and Europe, light, not water, is the most important limiting factor.

Many trees are deciduous, so evergreen understorey can be grown during the colder months. During the summer, the tree canopy shades most of the understorey and ground.

Food crop plants need to be chosen accordingly - they may need to be shade tolerant or exist and thrive in dappled sunlight.

In these countries the annual rainfall may only be 750 to 1000 mm (30 to 40 inches), but this is spread reasonably evenly over the whole year, while in places like southern Australia, most of the rainfall falls in the three or four months of winter, and the summers are much drier.

Another factor is that in the UK and Europe, evaporation doesn't often exceed precipitation. There is less need to build swales, dams and keyline draining systems because water is usually not in short supply.

Choosing a property

What do you look for in a property? Is it the gentle sun-facing slope, clear stream water or loamy soil? Different people value some things more than others and this will depend on the person's personal preferences and viewpoint. There may be times when you will have to assess different aspects of a property to see if it is suitable for your needs.

The following set of criteria, which can be used to assess a particular site, can be applied to rural properties of a few acres or more. You wouldn't consider many of these when buying an urban block, where scheme water is usually available and natural resources, energy supply and soil type are not that relevant. The criteria are not listed in any particular order or priority, but they are the kinds of things you can easily observe or note and assess. Any checklist for assessing a site might include:

- Buyer's resources: capital (money, machinery and so on) and their skills and abilities (including their willingness to have a go).
- Water: abundance and quality.
- Slope: this allows you to collect, store and save water.
- Aspect: sun-facing (northerly in southern hemisphere) for warmth, house site and better vegetation growth.
- Climate: annual rainfall, amount of frost, number of chill hours, length of growing season and temperature extremes.
- Soil type: clay, loam or sand.
- Natural resources: rocks, types of trees present and type of soil for

Figure 7.5 Natural resources are an asset on the property.

house building.

- Energy supply: available electricity or do-it-yourself solar or wind power generation.
- Altitude: more cold and frost in higher altitudes, microclimates present. Remember that slopes will often drain cold air.
- Access: roads, quality, need for repair, rebuilding.
- Cost and size: value for money, potential to increase in value, amount needed to be borrowed (and therefore amount needed to service the loan).
- Prior historical land use: traditional orchard that has been chemically sprayed or natural, untouched bushland.
- Amount of vegetation cover: trees, remnant bushland or totally cleared. Remnant bushland or forest should not be cleared for houses or other developments.
- Fences or hedges: type, quality, those that need replacing.

Figure 7.6 Find out about prior uses for the land.

- Views: aesthetically pleasing or will you be looking out over someone's untidy backyard?
- Activities of neighbours: traditional chemical farmers or responsible landowners and supportive community.
- Service facilities: public transport for school children, distance to shops.
- Fire danger risk: dry wind direction, flammable trees and plants nearby.
- Potential income: opportunities to produce income from resources on site or possibilities and potential for development, such as recycling grey-water, composting toilets and aquaculture license requirements.
- Suitable house sites: the selection of the house site depends on a number of factors, some of which may include slope, microclimate, water supply, soil type (stability, clay), drainage, fire risk and views.
- Closeness to markets: closeness for transport services - road, rail or boat - and markets to sell produce.

Some criteria are more important than others in different locations. For example,

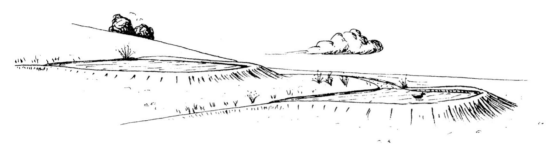

Figure 7.7 Potential aquaculture development.

developing aquaculture, forest industries and agricultural activities.
- Privacy: are there trees that screen you from neighbours or unwanted visitors. Can trees be planted for this purpose?
- Clearing needed: for house site, dam construction and garden areas.
- Drainage: does property flood or does water drain away into the soil?
- Dwellings on property: extent of useful sheds, cottages, and other structures such as a hothouse or shadehouse.
- Ecological value: natural ecosystems present or absent.
- Restrictions on land use: zoning, legal covenants, local or state government restrictions on the number of dwellings and on land and property

the activities of neighbours may be very important in one situation but less important in another, and energy supply may not be important if state-generated electricity and gas pass the property.

Remember that if you are contemplating buying a part of a farm or once rural property, the farmer or owner is probably selling the worst piece of land - that is usually subject to waterlogging, salinity problems and so on.

Make sure that you observe the land in both summer and winter, which will probably give you some indication about the property in these two extremes of climate and conditions.

Rather than try to prioritise the checklist of assessment for a site, it is better to allocate a value to each criterion depending on the property concerned. For example, water is usually the most important consideration (what grows without water?) and it might get a value of four or five out of five if abundant, clean water is available; only two or three if water is available only during winter and is stored in dams or there is a creek; and only one if you rely solely on annual rainwater.

If you set up a chart with these criteria you can easily assess them, allocate a score and add up the total to give a final assessment. One property can then be compared to another and decisions made about their suitability. You will not be able to find the perfect property, so analysis in some way like this is important.

If you are weighing up the pros and cons of a couple of properties, each of which has some good features and some not-so-good features, then analytically evaluating each site may be an option.

In this procedure, you simply assess each criterion about the block by a numbered scale. For example, in evaluating aspect on a site you might give it a ranking from 1 to 5, 1 meaning poor (shade-facing) and 5 meaning very good (sun-facing), with numbers 2, 3 and 4 meaning facing in other directions that you arbitrarily decide.

Each site or property is assessed in this way, and the one which achieves the highest total score is the best all-round property.

A further variation is to weight each criterion. For example, you decide that water availability is much more important than existing road access. Now you are considering the importance of some features over others.

You will need to prioritise the criteria and allocate a weighting for each. A simple example is given in the table below.

You may never find the perfect block - the one that has a gentle, sun-facing slope, great humic soil, natural woodland still present, the neighbours all using organic methods, plentiful, clean water, heaps of building material present (rocks, clay) and so on.

Realistically, we try to find a block with as many good features as possible and try to optimise the poorer features.

Local regulations

Before you or your client purchases a property, you should contact the local authorities about your plans. They will tell you what developments are allowable and what are not.

There may be restrictions on the use of land for particular purposes, or on clearing or about effluent and waste disposal. For example, you may not be able to run a school at a community you wish to set up unless the land is re-zoned for this purpose. Changing the zoning of land can be an expensive, lengthy and very slow process.

Furthermore, if naturally-occurring streams, rivers or waterways pass through the property, you should check with your

Table 7.1 Assessing properties. Property 1, with a higher overall score, may be more suitable for your needs than Property 2.

	Property 1			Property 2		
	weighting	value	total	weighting	value	total
plentiful, clean water	x10	4	40	x10	3	30
high humic soil	x6	2	12	x6	4	24
slope	x8	3	24	x8	3	24
sun-facing aspect	x8	4	32	x8	2	16
			total = 108			total = 94

local authority, or water authority or commission, about pumping rights and restrictions on water use. For example, you may not be able to dam the waterway, but you might be able to pump water to irrigate trees.

Similarly, there may be restrictions on house site and outbuilding placement. This depends on the property zoning, locality and bylaws of the local government authority.

These are the kinds of things that must be established before purchase of a property occurs. Your dream of building a demonstration permaculture site and working for a living on the land may not become a reality unless you've done some essential research about the site.

Finally, each property is unique, so the design for the site will also be unique. Wherever possible, try to incorporate unique aspects in the design. This will enhance the value of the property you have finally chosen and make it special.

My notes

Things I need to find out

8   Getting the house right: zone 0

The houses we design have to be harmonious with the environment, so where we live is an integral part of the holistic approach to the design of the property. It is said that houses are extensions of the people who live in them and that they reflect the beliefs and values of the owners. If this is true, and if we profess to be concerned about the Earth, then we ought to focus on making our houses simple, effective and efficient.

Zone 0 is the focus of a design and the house is usually the most important element in the design. Here we look at the principles of energy efficient housing and the integration of the house with the zone 1 garden areas.

The passive solar home

The energy needs of a home can be solved by a combination of design and materials. Good design permits passive solar gain, and building materials and hardware allow energy capture, storage and transmission.

A passive solar house is one which can effectively utilise the sun's energy to keep it warm or cool and also to moderate the adverse effects of climate on the house. Passive solar house design often includes:

- long axis east-west with the house twice as long as wide (or even longer), giving minimum exposure of the western wall to the sun.
- high percentage, 30 to 60% (depending on latitude) of glass on the sun-facing side.
- minimum windows on the east and west sides, especially the west which receives the severe hot summer sun.
- insulation on the prevailing wind and rain sides - usually the pole (sun-shaded) side of the house and either the west or east sides, or both.

Housing is relevant to the climate. For example, in temperate and Mediterranean climates, houses should be passive solar, with the long axis facing east-west (sun-facing), minimum windows on the prevailing cold wind sides, clerestory window banks for winter light gain and a

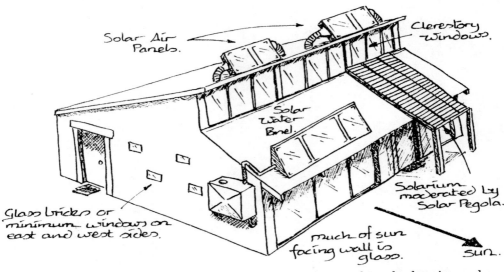

Figure 8.1 A typical solar house. The principles of a solar pergola and solar air panels are discussed under the heading 'retrofitting' later in this chapter.

solarium or conservatory on the sun-facing side for further winter warmth.

Houses in tropical, subtropical and desert areas should be designed so that they remain cool during the summer day and warm during the summer and winter nights. Desert areas are well known for their large temperature extremes - a range of 30 to 40°C in one day is not uncommon.

These types of houses need to be shaded (for example, by plants on a trellis), cool winds need to be directed toward the house, and surfaces painted white for light reflection. Underground houses or earth-covered shelters are useful building strategies.

Underground and earth-covered houses are remarkable in that the temperature, both day and night and season to season, is fairly constant and within the human comfort level at all times.

Another building strategy for a hot climate is to draw cool air through the ground into rooms. For example, cool air from a cellar can be moved (by convection or fan) to various parts of the house, as shown in Figure 8.3, or air can be passed over a wet surface or pond as a form of evaporative cooling.

the house into the home. A typical Australian "outback" homestead, for example, is a house with verandahs all around, a silver-coloured tin roof for light reflection, and outdoor cooking facilities.

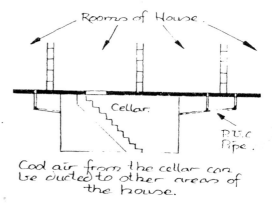

Figure 8.3 Large diameter PVC pipe is used to allow air to move from a cool cellar into warmer rooms.

The secret in keeping a house warm is to reduce heat loss. Strategies to maintain heat or minimise heat loss in a house include insulation, double glazing (especially on the sun-shaded side and in the direction of cold winter winds), increasing

Figure 8.2 An example of an earth-covered house with a sun-facing side exposed.

Strategies to cool the house include insulating the roof and walls, building verandahs or porches and pergola at least on the west side, planting deciduous vines and trees, having breezeways which direct cool winds, and drawing cool air from a shadehouse on the sun-shaded side of

the thermal mass of house (brick walls, dark slate on floor), increasing the percentage of glass on the sun-facing side (north in southern hemisphere), sealing gaps under doors and around loosely-fitted windows, and building a hothouse or greenhouse on the sun-facing side.

Sod, or soil covered, roofs may be another

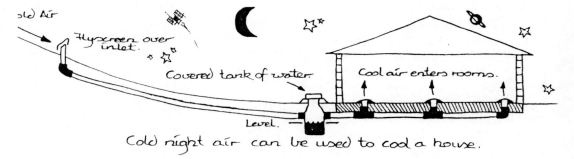

Figure 8.4 Cold night air is cooled as it passes over water before it is ducted into the house.

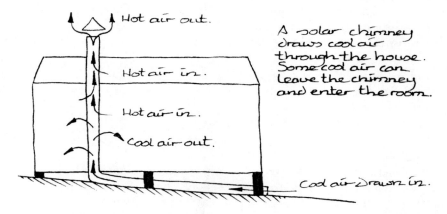

Figure 8.5 A solar chimney is one way to cool the interior of a house. Air is drawn from below the house, often across water, and ducted through the house and then vented.

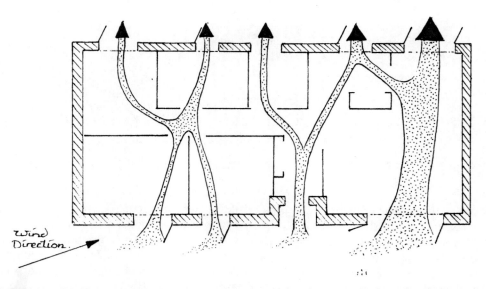

Figure 8.6 Directing breezes through a house by efficient cross-ventilation is an important strategy in regulating the house temperature.

strategy that can be used to insulate a building. The only two considerations are the damp-proofing and the extra roof weight. Thick plastic sheeting is laid over the roof before 75 to 100 mm of soil is added. The sheeting allows rain water to drain downwards towards the gutter. Extra re-inforcing timbers have to be placed in the roof to support the additional weight. You could expect about 150 kg or more of wet soil per square metre of roof area (100 mm thick).

Plants, such as grasses and herbs, can be planted on top of the sloped sides and roof for further heat control and soil stabilisation.

For houses in warm climate areas, it is not important to orientate the house toward the sun for passive solar gain. Houses need to exploit cooling breezes and increase wind movement (ventilation) through the house.

Evergreen trees for shading and shelter for the house from potentially strong winds are important. In tropical areas, where hurricanes and cyclones are a threat, re-inforced dwellings, sometimes raised off the ground, are built.

In many climates solar hot water systems should be mandatory. These can provide free hot water for most of the year, and when it is raining or cloudy, various booster systems can be employed These range from solid fuel to gas to electric boosters.

There are two main types of solar hot water systems. In a passive hot water system, which relies solely on natural convection currents to move water, the storage tank must be slightly higher (300 mm is adequate) than the panel so that cold water from the tank falls towards the bottom of the panel. The collector panel heats the water and the hot water rises towards, and into, the storage tank.

In an active system, energy is used to pump water through the panels to and from the storage tank. The tank can be positioned anywhere, such as in the roof space above the bathroom or outside on the ground near the kitchen. Sensors detect the temperature differences between the panel and tank and, usually when the panel is hotter than the tank, the pump is engaged and water starts to flow.

Integrating the house and garden

Ideally, the house and garden should complement each other. The garden can be designed to take an active role in the heating and cooling of the house. For example, deciduous trees should be planted along the sun-facing wall of the house. These regulate microclimate by providing shade in summer and allowing sunlight to filter through during winter.

There should be a mixture of deciduous and evergreen trees around the house. A sector plan of sun angles will show you where each type of tree should be planted. Evergreens can be planted in areas around the house where only summer sun reaches, such as the eastern and western walls. The winter sun angle for your property might be narrow, and only deciduous trees should be planted in this sector.

As these trees are found in zone 1 you

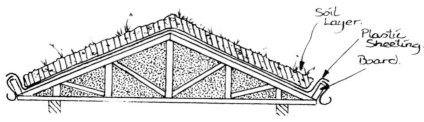

Figure 8.7 Extra timber beams are needed to support the additional weight of a sod roof.

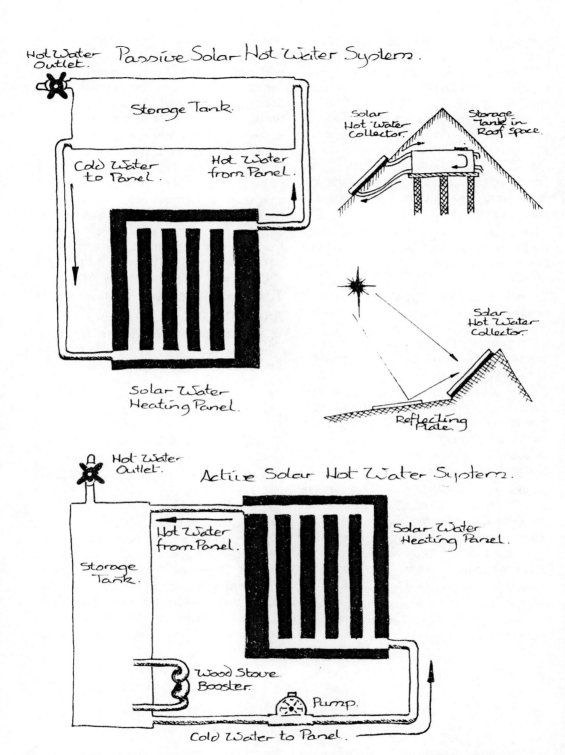

Figure 8.8 The differences between a passive and an active solar hot water system. Also shown is a passive system with the storage tank in the roof space and one way to increase energy gain, by using a reflecting plate to direct light and heat waves to a solar water collector.

Figure 8.9 Deciduous trees and light coloured paths both contribute to the amount of light energy entering a home during winter.

should consider planting evergreen fruit trees such as citrus and deciduous fruit trees such as stone fruit.

Garden structures can also be used for temperature control. A hothouse attached to the main house can supply additional winter heating for the home as well as being used to grow food during the cooler months. A hothouse works when light penetrates glass and is absorbed by objects. Heat energy is re-radiated as longer wavelengths which cannot pass through the glass layer. This hothouse, or greenhouse, slowly heats up.

By placing a few windows or vents between the house wall and the hothouse structure, warm air can be drawn into the house. Examine Figure 8.12. During winter the external vent is closed and the internal windows or vents between the house and hothouse are open. Warm air produced during the day can be directed into the house. Cool air from the house will enter the hothouse to complete the convection cycle of air movement. A word of caution: if you use water in the hothouse to grow plants, the high level of humidity may cause moulds and mildew

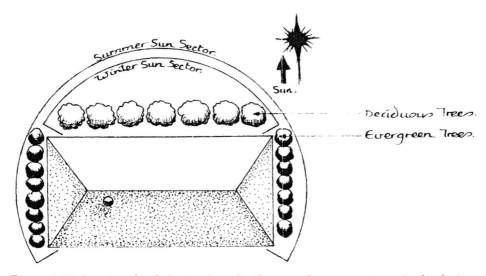

Figure 8.10 A sector plan helps to place deciduous and evergreen trees in the design.

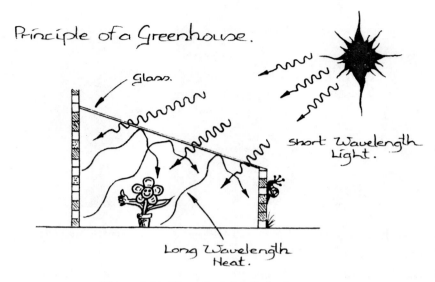

Figure 8.11 The principle of a hothouse.

to grow on walls. You need to vent well and keep the hothouse as dry as possible to minimise this problem.

During summer, the internal vents are closed and the external vent is open. Hothouses get very hot in summer and are seldom used to grow food during these months. The hot air is ducted to the outside atmosphere.

Radiant heat can also be used to help plants growing near the house. You can grow cold-sensitive plants near tanks of water or alongside stone or earth walls. Water has a high heat capacity and slowly radiates or releases its stored heat as temperatures fall. Frost-sensitive plants are kept a little warmer in this way, which is an excellent example of the use of microclimate in design.

Heat is also available from other sources, and can be directed from the house into the garden. The major hot greywater source is from the bathroom - either the shower or bath.

Only a small amount comes from the kitchen sink. If people consistently use hot water for washing clothes, then this source also needs to be considered. Hot

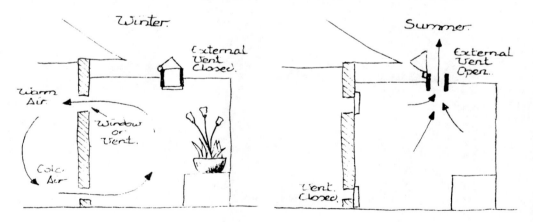

Figure 8.12 A hothouse can provide additional winter heating for a home.

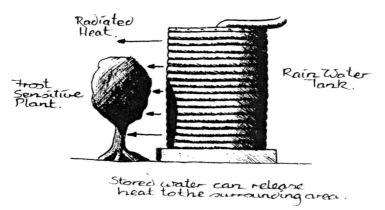

Figure 8.13 A rainwater tank can keep plants from freezing. Rainwater tank overflow can be directed to a wet, bog, and/or shaded garden, where many food plants can be grown, including taro, watercress and sweet potato.

water could be isolated from the other greywater sources and directed through the floor of a hothouse so that this heat can be used, especially in winter to warm the surroundings.

Garden microclimates are also created by areas of shade and coolness. The house and attached structures can provide the necessary shade and coolness required by some types of garden plants.

Structures such as trellis can be built horizontally over the garden for hot, arid climates, but it is mainly positioned vertically for maximum sunlight and heat gain in the colder climates. It is common to see vertical trellis supporting peas, beans and berry fruits, such as loganberry and boysenberry, in colder areas.

Trellis and pergolas enable humans to also enjoy the garden in the hot months. These structures can provide quiet, shaded areas in the summer garden.

A variation of garden trellis is the vertical living screen. Many west-facing house walls absorb considerable heat during summer. These walls can be protected by building a mesh screen or vertical trellis structure close to the house wall. Shading

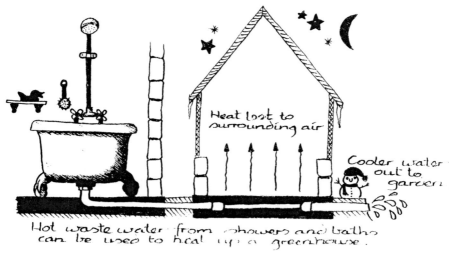

Figure 8.14 Using waste heat from showers and baths can supplement the heating of a hothouse.

Trellis protects vegetables in hot climates.

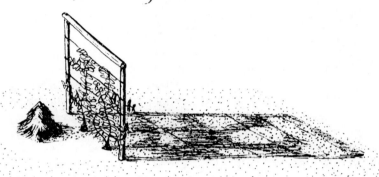
Vertical trellis is a good way of gaining heat in cold climates.

Figure 8.15 Trellis protects the vegetable garden in a hot climate.

by evergreen creepers will lower the heat gain by the wall, keeping the house cooler in summer.

The screen can be built ten to twenty centimetres from the wall so that a layer of air is trapped between the screen and wall, further insulating the wall from heat radiation. Alternatively, place the screen a metre from the wall and make a sheltered pathway. Either way, protecting walls from direct heat radiation will lower the overall temperature of the house.

Consider children when you are designing the house and garden. Place a play area where children can be seen and supervised from the lounge room, kitchen or wherever adults tend to be at the time. The play or garden areas should be just extensions of the house and just another area where children can learn about and experience life.

The extent and nature of play areas depends on the ages of children or grandchildren. Young children like to hide, while larger children (adolescents) tend to throw or kick balls about, or shoot a few basketballs.

Views overlooking play areas also often have a focus such as a pleasant looking building, a seat, a pond or a particular tree.

Retrofitting existing houses

In the world in which we all live, owning a house, let alone building one that you would like, is not a reality for many people. Some people do own, or are paying off, their houses, but many people cannot afford this luxury.

You may have to examine how the design of buildings on a property (which is just as important as the design of the gardens)

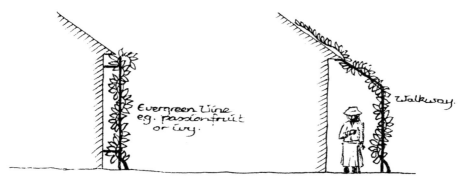

Figure 8.16 Air is a good insulator. A vertical living screen will shade a wall and reduce heat gain by the house during summer.

fits into your particular situation. In most cases, retrofitting an existing house is far more economical than pulling it down and starting again.

Retrofitting is the term that describes the changes you make to an existing energy-inefficient house so that it becomes more energy efficient.

In many cases, the house utilises passive solar gain and design strategies to minimise heat loss and maximise heat gain during winter and vice versa for the hotter summer months.

You may be able to suggest retrofitting ideas to ameliorate the influence of climate on the house. These include the use of shadehouses, greenhouses, trellis and pergolas, earth berms, the placement of deciduous and evergreen trees, windbreaks and ponds.

Some of these were discussed in the previous section, so we will focus on three simple ideas as examples of retrofitting strategies.

A solar pergola is an effective way to allow winter sunlight into a house but keep summer sunlight out.

Wooden or sheet metal blades are placed at an angle so that summer sunlight cannot pass through, but winter sunlight, at a lower angle, can.

The blades are normally fixed and set, so that maximum light will pass through into a room at the winter solstice, when the sun is at the lowest altitude in the sky.

Spacing between the blades regulates when sunlight is able to pass through, so that you could have shading for four or more months during the hottest part of the year.

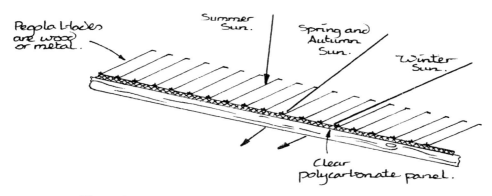

Figure 8.17 A solar pergola allows winter sunlight into a house.

In existing houses you may have to replace some of the roof panels with clear polycarbonate sheeting or glass so that sunlight can enter a room. The solar pergola is placed and supported on top of this. A dark slate floor or walls with high thermal mass will absorb this energy and release it during the night.

Solar air panels can be easily made and fixed to a roof. These are similar to solar hot water system panels except that air, and not water, is passed or pumped through them.

During the winter months air is drawn from either rooms, roof space or the outside atmosphere and is heated as it blows across the panel. The hot air is then ducted into the house and can enter designated rooms. Sensors detect differences between room and panel or outside environmental temperatures, causing an air fan or pump to turn on or off. The movement of air is thus regulated. Naturally, the solar air system doesn't operate during summer.

Most heat loss from a house occurs through the roof and roof space. Insulating is an obvious solution and this does indeed help keep the house cool in summer and warmer in winter. However, you also might want to vent the roof space so that the hot air that builds up in the summertime can be directed away into the atmosphere. Even drawing hot air from rooms via the roof space is possible as shown in the following diagrams.

There are many other ideas about retrofitting. These include placing pelmets and ceiling-to-floor curtains over windows to reduce convection currents, blocking off windows on the western side and building windows on the sun-facing (equatorial) side, placing door stops and seals underneath and/or behind doors to reduce wind flow-through in winter and subsequent heat loss, and building sheltered skylights in the roof which only allow additional

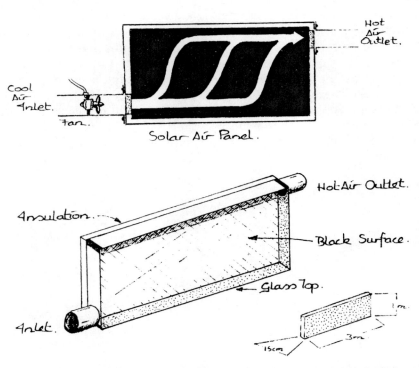

Figure 8.18 A solar air panel heats up cold air and blows it into the house.

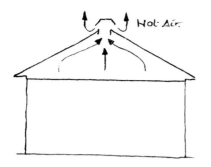 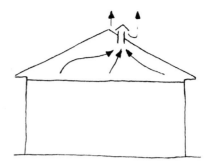

Figure 8.19 Roof space can be vented for summer cooling and winter warmth.

winter sunlight into the house. The number of retrofitting strategies are endless and I'm sure you'll think, or come across, lots of other ideas about how to make your home more comfortable and more energy efficient.

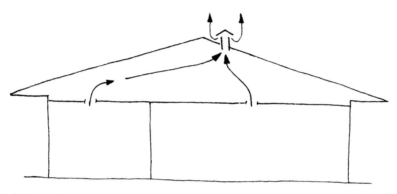

Figure 8.20 Warm air can be ducted from the ceiling area via the roof space to the outside atmosphere during summer. Vents are closed during winter.

My notes

Things I need to find out

9 Water harvesting

Pure, clean water is the number one priority for any permaculture system. Permaculture designs try to harvest, retain and reuse as much water as possible before it is lost from the system. This is especially true in dry and arid climates where rain mainly falls in a few months. In countries such as the UK and those in Europe, rainfall is often spread over the whole year and water harvesting techniques are not so critical.

In drier countries we need to incorporate a range of strategies to harvest and store as much water as possible. Think of the soil and your garden plants as water storage vessels. A forest of trees is sometimes referred to as a lake above the ground because large amounts of water can be found in the living tissue of every plant.

On the suburban block

Strategies to harvest water on urban sites include directing rainwater from the roofs of houses and outbuildings into the garden, swales in sloping ground, drains, ponds, small dams, rainwater tanks, reusing greywater onto the garden (check first to see if this is illegal in your area), wells and bores.

To calculate how much water falls on your house roof, multiply the average yearly rainfall by the roof area of the house. For example, for a roof area = 200 m^2 and annual rainfall = 1000 mm or 1 m, 200 m^3 or 200 000L can be collected (200 x 1000). This is quite a lot of water, provided you can harvest and store all of it.

Gutters are not needed on houses unless rainwater is collected and stored. A stone or pebble filled trench is used to drain some water away and to direct water into the nearby soil.

Terracing is another way to build gardens on slopes. The soil is removed in a cut-and-fill operation. You usually have to

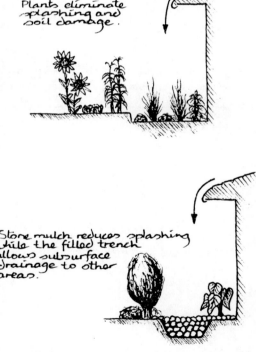

Figure 9.1 Rainwater collected from a house roof can be stored in a tank.

Figure 9.2 Rainwater can be directed into the garden.

build a retaining wall. This can be made from tyres (filled with soil), sheets of roofing iron or wooden planks - whatever you can afford or what are available as inexpensive resources. The terrace should be wide enough to include a path and garden bed - up to two and a half metres across, with half a metre for the path and up to two metres for the bed. Remember, the bed can be this wide because you have access to both sides - although you only have to stand on one side but bend over the other.

Greywater is another water harvesting strategy to provide additional garden water for your home. Greywater is the domestic wastewater. Before you decide to build a system such as that shown in Figure 9.4, check with your state or local government authority to see if you can install this type of system. You may be able to direct all of your bath and laundry water into a reed bed which will remove many of the nutrients, disease organisms and pollutants before you pour it onto your garden.

It is important to use local wetland plants. These types of plants are especially adapted to exist in waterlogged soil. Their leaves can transport oxygen from the air to their roots so that all parts of the plant can function effectively. Small amounts of

Figure 9.3 Water will flow down the slope into garden beds.

wastewater from sources such as the kitchen sink, the shower and bath, handbasins and laundry trough. If the toilet water, commonly referred to as black water, is isolated and treated separately, then the greywater can be recycled on site.

Water which you wish to reuse can be passed through a biological filter. These can be reed beds of plants which are able to take up excess nutrients and store them in their tissues.

Reed bed systems are now being used throughout the world to treat and purify oxygen diffuse out of the roots into the surrounding water medium. This oxygen supports a large range of bacteria and other micro-organisms which are important in the treatment of wastewater.

The most commonly-used reed bed plants are species of *Typha* (bulrush) and *Phragmites* (common reed) but these produce wind-blown seeds and their use should be limited.

Contact your local university, herbarium or wetland management group to find out what plants are in your area and if they are suitable.

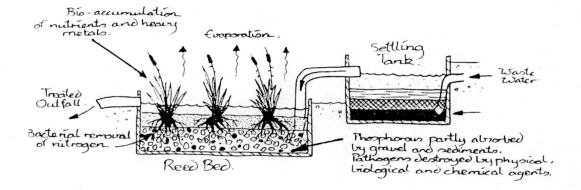

Figure 9.4 A typical reed bed system to treat and recycle your greywater.

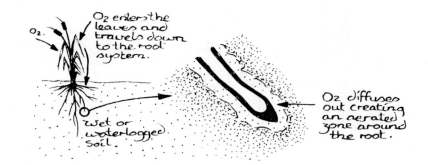

Figure 9.5 Wetland plants can transfer oxygen from the air to the plant roots.

On the farm

Water should be stored in the soil. Every effort should be made to get water into the ground. It is important to reduce water flow (slow it down) and hold it; otherwise erosion could occur. For small-acre and broad-acre properties water should be stored as high as possible. Provided that water can be economically placed there, dams can be built on ridges, on top of hills and at the keypoint.

Keyline cultivation

Before we launch into the concepts of keyline cultivation, we need to discuss contours and their relevance in holding, storing or moving water. A contour is an imaginary horizontal line, where all points along it are at the same height or altitude.

A contour line is at right angles to the slope of the land, so as slope changes so too will the contour lines curve and turn.

A contour map displays the location of contours, and landscape and many topographical details can be interpreted from it. For example, lines close together suggest steep slope, contour lines further apart indicate flatter, valley ground, while circular contour lines suggest hills.

Contour lines are arbitrarily placed at 5 or 10 m altitude intervals, so one line will be 5 or 10 metres above or below the next line. From the number, spacing and shape of these lines on a property, you can determine the total amount of fall, the degree of

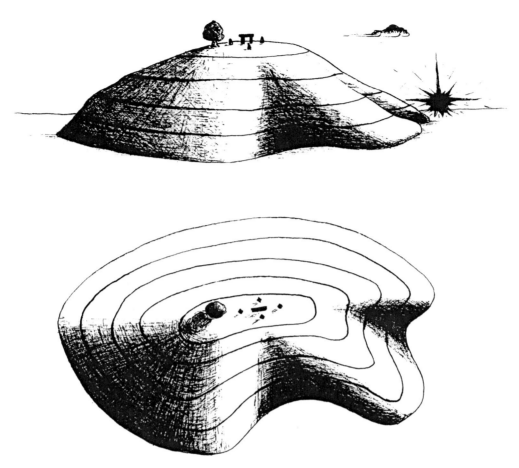

Figure 9.6 A contour map shows the amount of slope and landform change on a site.

slope and the possible location sites for dams and other water harvesting strategies.

The keypoint is identified by finding the start of a valley. This can be found by noticing either a slope or contour change somewhere on the hillside. You can often find evidence of the valley forming in the upper half of the hillside, or somewhere around the middle. This can be confirmed by measuring the contours at and around this point.

The point at which the slope changes direction, from convex to concave, is called the keypoint. The keyline is the contour at this point. Other lines are drawn parallel to the keyline as part of keyline cultivation. Some of these keylines will be off the contour, permitting water to flow in the direction towards the ridges.

Dams are constructed along these keylines so that they are linked and water can flow from one to another. A dam at the keypoint, or start of the valley, would be one of the first and highest in the slope. Generally, dams built in the highest country would give the greatest potential benefit, as water can be easily stored and moved. Ideally, dams are dug and built on poorly drained sites, such as clay areas. Consideration must also be given to the storage ratio. In other words, how much water can be stored for the amount of soil that is moved to construct the dam. A

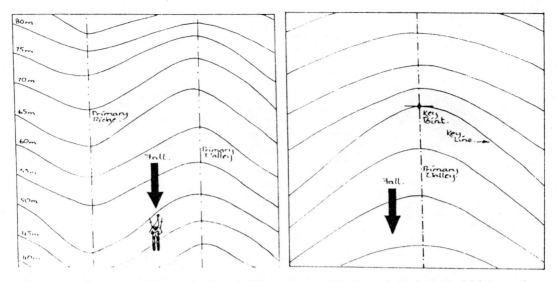

Figure 9.7 Contours, ridges and valleys in the landscape. The keypoint is found by looking at the changes in slope.

ratio of 3:1, or higher, is best and this ratio generally increases with decreasing slope as the land flattens.

Slope on your property is to be used for advantage. Water that is stored higher up the slope can be let to irrigate your garden area by gravity.

Water can also be directed from the valleys or gully areas, where it accumulates, to the drier ridge areas by making it flow along contours as part of the keyline system of drains and channels. Generally, steeper ground is drier and shaded areas are wetter or, at least, dry out slowly.

To calculate how much water you can expect to harvest you have to find out the annual rainfall for your area or property. Imagine if you owned, or were doing a design for a client who lived on, 400 ha (1000 acres).

For a 1000 mm annual rainfall and 10% water loss, you would expect 4 000 000 m^3 or 4 000 million litres harvested, of which 400 000 m^3 or 400 million litres would be lost each year. To visualise a 1000 mm rainfall imagine water one metre deep completely over the land.

Keyline cultivation is like hundreds of small absorbent drains, each holding some water, so that the water does not fall towards the valley. Keyline cultivation is primarily a soil improvement system.

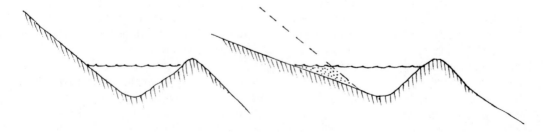

Figure 9.8 The higher the slope, the lower the storage ratio. The flatter the slope, the more water that can be stored efficiently.

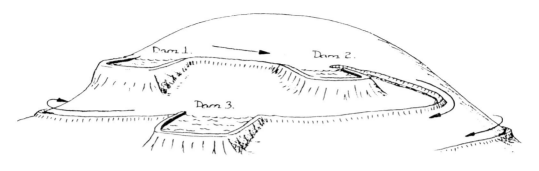

Figure 9.9 Dams should be built along a slope so water can move by gravity.

As soil becomes fertile, more water can be absorbed and stored and the amount of air in the soil increases.

Along with moving water out towards the ridges is the replanting of trees in these areas. Trees should be left or replanted on all ridgelines.

Furthermore, if ridge areas contain remnant vegetation, then these areas should be fenced to prevent stock from destroying what is there.

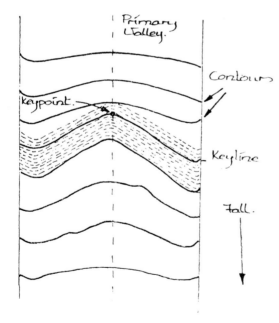

Figure 9.10 Keyline cultivation makes furrows along lines which are parallel to the keyline.

Dams

When siting a dam use the keyline strategy. Don't put the dam in the valley in the belief that the bottom is best for water collection. Place smaller dams higher up the hill at the start of the gully or valley, to harvest water which can be stored and then moved by gravity alone.

Remember to scalp the topsoil and store it when building a dam or doing major earthworks. The topsoil (the first 15 cm or so) should be used to help rehabilitate and landscape the changed area, so that plants and soil systems, such as fungi, bacteria and animals, become re-established. However, do not leave a large heap of top soil for too long - the soil life will eventually die.

There are many types of plants which can assist in stabilising dam walls. Bamboos and clumping grass species, such as vetiver grass and banna grass, are useful.

You can also sow oats, native grasses or living mulch plants. These will hold the soil and increase the humus and nutrient content of the soil, if they are slashed and left as a surface mulch. Large trees should only be planted beyond the walls as they often fall and remove the dam wall as they do.

Dam wall slopes should be at least 1:2 and preferably 1:3 or 1:4. This is the ratio of the height of the wall compared to the base. Dam walls will slump and fall in if they are too steep.

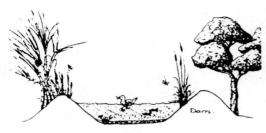

Grasses, bamboos and small trees protect water supplies and help bind the soil.

Figure 9.11 Large trees can shade and protect water supplies in dams. Grasses, bamboos and ground covers are useful plants for slopes as their extensive root systems help bind and protect the soil.

Occasionally, a silt or chemical trap may need to be built whenever severe run-off is known to contain silt, fine clay or excess nutrients.

Wetland plants can be used to help filter and settle soil particles from the water as shown in Figure 9.13.

This holding pond can also be used as an initial treatment site for fertiliser accumulation. For example, excess phosphate can be precipitated by adding alum, and calcium and phosphate can be removed by adding soluble ferrous sulphate where the sulphate links with calcium and the iron (ferrous) with phosphate.

Some nitrate will be removed by uptake from the aquatic plants in the pond. You also may need to consider possible health problems, such as bacterial diseases, caused by contamination of the soil or water by animal droppings.

Drains which collect water can have several uses. The structure and construction of these drains vary depending on the function you wish them to perform. For example, drains can be called swales, diversions, interceptors or spreaders depending on the situation.

Moving water through drains

Swales are ditches on the contour. When they fill or catch water, the water does not flow away. They hold run-off water and allow it to seep into the soil. Furthermore, placing access roads along contours in effect doubles up as a swale where water can be collected and/or directed to other parts of the property.

Swales work best on sandy soil because the water can be caught and then allowed to soak into the ground. If swales are constructed in heavy clay soil you may have to rip the ground and/or add gypsum which will coagulate (clump together) the clay particles. This will improve drainage, and water and root penetration of the soil.

Deep ripping of hard ground to about one metre is often used before tree planting occurs. Only that area where trees are actually planted needs to be ripped.

Small swales in suburban backyards or on small acreage properties can be mulched and sown with seeds to grow food crops such as melons. You don't normally follow this practice on rural properties because of the expense and time needed for the operation. However, aim to utilise all parts of the swale.

Dam walls need to be sloped.

Figure 9.12 The slope of dam walls should be such that they don't cave in. Slopes of 1:2 (27°) or 1:3 (18°) are the minimum.

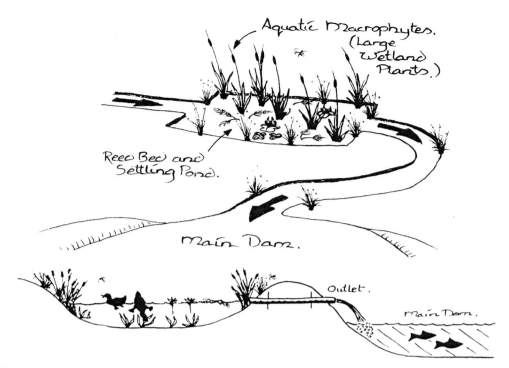

Figure 9.13 A settling pond and reed bed is used to remove excess nutrients. Here are two different ways this can be set up.

Figure 9.14 A swale is a ditch cut along a contour. Water is held in the ditch until it soaks into the soil.

Figure 9.15 Permaculture uses multifunctional elements. An access road built along a contour can also act as a swale.

Swales also trap organic matter and the ditch becomes a rich, thick layer of humus which holds lots of water. You can plant directly into the ditch (good for melons) or use the ditch as a pathway. Eventually, the swale will fill up and you might end up with a terrace.

Terracing is possible on some slopes, but is more common in tropical and subtropical areas than in dry and Mediterranean areas.

Terracing is effective in high rainfall areas as a means of harvesting water, reducing erosion and growing water-loving plants such as taro and rice. Terraced slopes still need a range of plants to stabilise the soil.

Swales in farm paddocks will naturally accumulate animal manures from stock, thus providing a nutrient-rich fertiliser for trees grown in or on the swales. Swales should be wide enough to permit access - either footpath, wheelbarrow or even tractor.

Flood irrigation can be achieved by using spreader drains, which are essentially like swales in that they are level and on the contour. Water from a dam or stream is directed into the drain, it fills up and overflows over the top side, along its length, and cascades downhill over crops or paddocks. A word of caution: on steep slopes, and compounded by non-wetting soils, sheet erosion can occur. Spreader drains are only suitable for gentle slopes.

Unlike swales, which are normally built on permeable soils, diversion drains have a slight gradient. Even a slope of 1:1000 will drain water, but gradients of 1:100 or less are much more effective. This means that there is a one metre fall over a distance of 100 metres. A slope of 1 in 200 is

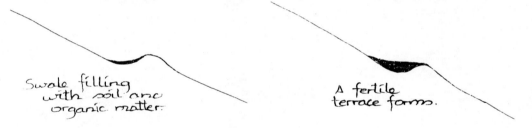

Figure 9.16 Terraces form when swales fill up.

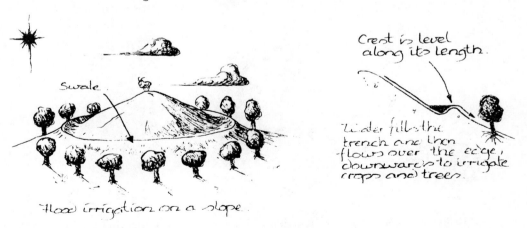

Figure 9.17 Spreader drains can be filled and then allowed to overflow to irrigate crops.

enough to move water and slow enough to minimise erosion and unwanted movement of sand and silt.

Diversion drains work better when the base and sides are clay-lined. This may mean that the drain cut is deep enough to slice into the clay layer, often below the sand layer. Drains may be 0.5 m or more deep.

Interceptor banks, based on the work of Harry Whittington in Western Australia, are known as WISALTS - Whittington Interceptor Sustainable Agricultural Land Treatment System.

These types of drains are always cut into the clay and compacted, no matter how deep it is (often up to 2 m down).

The clay is used to line the trenches, which can be cut on the contour, as a swale is, or cut slightly off the contour so that water drains away. If no clay is available, plastic sheeting may be used.

WISALT drains are used to take water and salt-affected water away from waterlogged areas, such as paddocks, and discharge it into dams or streams.

The compacted, clay-lined, downward side prevents salt-laden water moving through the soil downwards, thus reducing further soil collapse.

In effect, interceptor banks drain and dry the soil, removing potential waterlogging problems, including problems of anaerobic conditions and plant death.

WISALT banks should only be installed by licensed consultants, because the spacing and placement of these types of drains is critical to the success of the system. These trained consultants can recognise natural field barriers, dykes, throughflows and recharge areas. They also understand the chemical changes that take place under waterlogged conditions, as well as perched and rising water tables and other water movements in the landscape.

We have to address the salt and waterlogging problems by drainage, soil improvement and so on before trees are planted.

Degraded land and soils, with high salinity problems, arise from either groundwater rising and bringing dissolved salt to the surface or from soil collapse, where the land is cleared, cultivated and compacted, and the soil structure breaks down or changes.

Erosion removes the topsoil. There is less humus and air, and more waterlogging. Clay seals the ground on the surface and often somewhere below the surface. In effect, a soil swamp or bog is produced in this collapsed soil situation.

Usually before soil collapse occurs, there are tell-tale signs, which include particular vegetation growing in particular areas (rising water table, waterlogged areas), such as reeds, barley grass and dock, or a noticeable wet area which remains wet well into the summer (farm vehicles easily bogged). Some salt is usually evident

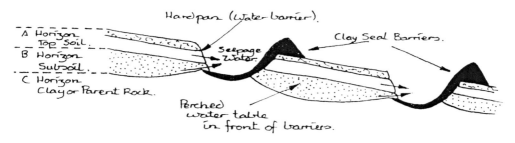

Figure 9.18 Interceptor banks have to be cut into the clay layer to prevent water from seeping downslope. The water can be used upslope - it moves through the soil by capillary action - or taken to a dam or safe disposal site.

on top of the ground. Generally, there is also evidence of reduced crop yield or pasture growth, as well as the yellowing of plants in waterlogged areas.

You can use a backhoe to dig a three metre deep test hole which will allow you to determine if soil collapse is occurring. Water that trickles or seeps from the top layers downward to fill the bottom (diagram A below) suggests soil collapse and a hardpan clay layer, causing seepage. Water that appears to well upwards from the bottom of the hole (diagram B) also can suggest poor drainage. The digging releases pressure and the water bubbles upwards. This may also indicate possible salt problems due to the collapsed soil.

Strategies for both of these situations would include building interceptor banks to restrict the recharge water from mov-

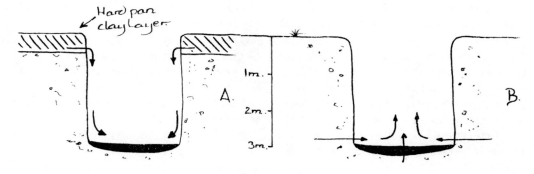

Figure 9.19 Digging test holes to examine soil collapse and drainage.

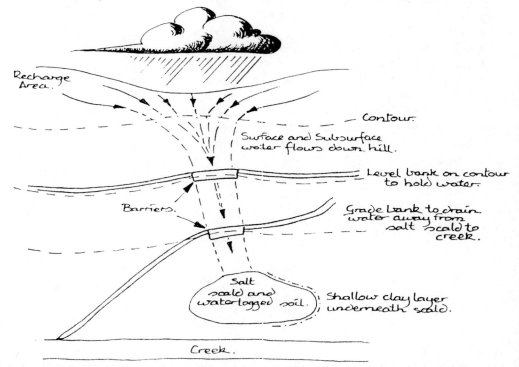

Figure 9.20 Clay-lined drains are barriers to water movement downslope. These drains can divert water from salt scalds and help to dry paddocks which waterlog easily.

ing through the soil. Planting trees in these recharge areas would also help to utilise some of the excess water. You should expect that good soils, with a high holding capacity, will have little water movement through them.

Land clearing has also caused the water table to rise, bringing salt with it to the surface. Trees naturally act as pumps, keeping the water table below the surface. When these trees were removed the water rose.

Waterlogging and flooded areas killed all soil life as the air was expelled. The soil structure collapsed. Soon, all you see are ever-increasing areas of salt scalds. Some salt is brought up from lower soil layers, but much of it is trapped above the hard pan and cannot drain away. The salt slowly accumulates and the concentration increases. Bacteria and soil organisms die,

below the surface. The soil becomes waterlogged. When this happens further rain cannot enter the soil at all and more overland flow occurs. The problems escalate.

Keyline cultivation is used to break up the hard pan, allowing water and air to enter and infiltrate the soil. Water is then permitted to permeate deeper into the soil,

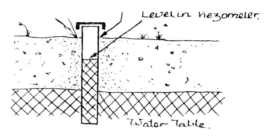

Figure 9.22 Piezometers allow you to examine water table height and water pressure and salinity level.

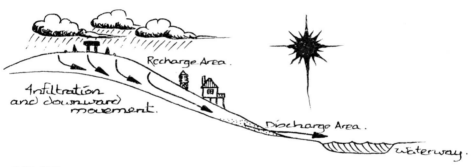

Figure 9.21 Hills are often recharge areas where rainwater enters the soil and moves down into the soil and down the slope.

the soil becomes anaerobic and stock refuse to eat the remaining grass growing on it. After land is cleared some soil, the topsoil, is often washed away by erosion. Small soil particles are also washed downwards where they compact and cement together to seal the lower subsoil.

This is what is called the clay hardpan, which gradually rises as more topsoil is lost and greater compaction occurs.

As rain falls, most is lost as run-off and what does enter the soil permeates only to the hard pan, which may be only 200 mm

where it is held, and thus able to be used by plants.

Planting trees often doesn't solve the problem - even if some of these plants can survive in waterlogged soil. The environmental movement's push to "green" the world by planting trees is futile if the soil is not suitable for the trees to grow.

Some interceptor banks hold the water where it is needed most - on the slopes. The trenches don't allow water to drain away, carrying nutrients, humus and organic matter with it. These materials stay so that plants can access them.

Banks by themselves are of little use. Trees have to be planted along the whole length of the drain. Four or five rows of trees are not uncommon and you could expect to plant about 1000 trees for one kilometre of drain.

If stock are to be kept in areas containing drains then fencing of the trees must occur. Even electric fencing, which is the least expensive option, does increase the cost of land management strategies.

However, unless fences are erected stock will quickly devour the shrubs and trees. In time, stock can be rotated through the fenced tree belts as part of the farm grazing system.

Swales and interceptor banks are spaced depending on soil type, rainfall and slope. Usually swales are close together on the steeper slopes and further apart on the shallower gradients. You may, for example, cut swales every hundred metres on one part of the farm and only half that distance on a steeper area. However, machinery cannot operate safely on slopes greater than 20°. On slopes of this size, or less, a bulldozer may be able to push the soil into a bank as it moves downwards. This is the best way to build the drains.

In dry areas

In dryland areas it is better to store the water underground. In hot climates, water quickly evaporates from dams, so it is best to make water fall into trenches and pits, where it can infiltrate into the nearby soil and supply nourishment to the root systems of plants.

Terraces can also be used in drier areas. Small patches of cereal crops or green manure crops, such as mustard, can be easily grown and harvested.

Terraced slopes are normally cut by hand - using tools of course. Planting on slopes is essential for soil stability. Vetiver grass, *Vetiveria zizanioides*, is a remarkable

Figure 9.23 When building swales and interceptor banks, start at the top of the hill and work downwards.

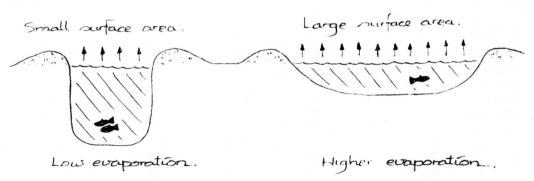

Figure 9.24 Dams should be deeper in drier areas to minimise evaporation.

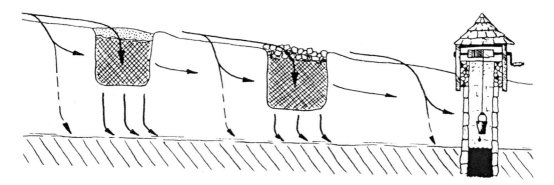

Figure 9.25 A pit filled with compost and covered in stone or sand can trap and store water. Natural seepage and drainage of water will slowly fill pits and wells.

plant that has a very long root system and can survive in both drought and flood. Hedges of vetiver grass are used in arid areas for erosion control and can trap and accumulate large amounts of topsoil, thus creating mini-terraces at the same time. What's more, it's good fodder for stock and easy to propagate. Lemon grass, *Cymbopogon citratus*, is also good for erosion control.

Water use in the garden should be monitored and controlled. Your aim is to obtain the optimum benefit from the water that is used and to minimise evaporation and seepage away from plant root zones.

Bury unglazed clay pots in the vegetable garden, fill with water and leave it to slowly permeate through the walls into the surrounding soil.

These types of strategies should be tried more in the drier areas of Australia and America, as at present they are used mainly in Africa.

In dry areas, the daytime soil temperatures can be very high. The surface could

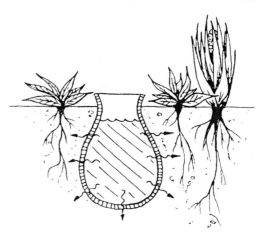

Figure 9.27 Water held in clay pots can diffuse into the surrounding soil.

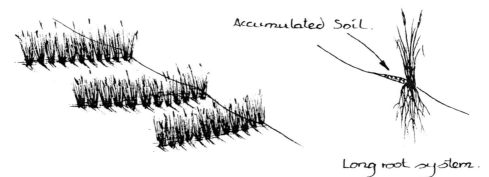

Figure 9.26 Vetiver grass is used to reduce erosion on slopes.

be over 40°C. These high temperatures will cause stress to plants, especially exotic garden vegetables, and high transpiration rates resulting in wilting and possible death.

The use of shade trees, including some that are nitrogen-fixing and general soil builders, over a food crop is common in some countries.

Mulches should be used in these areas too, as these will help to keep the soil cool and retain moisture. Even stone mulches can be used when this resource is available. Often, timber-based mulches are just not available or would be expensive to buy and transport in.

As water is very scarce in arid climates many strategies must be tried, even wire fences which allow dew to condense and drip onto the soil. Dew is a limited water collection strategy, but in these areas any water is helpful for plant survival.

Figure 9.28 Even a stone mulch will protect and cool the soil.

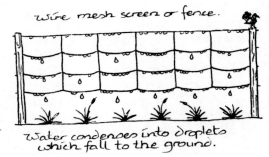

Figure 9.29 Dew will condense on objects. Water can then drip into the soil.

My notes

Things I need to find out

10 Designs for urban settlement

Permaculture designs for suburban yards and small city properties have to primarily consider the lack of space and the lack of natural vegetation cover. The suburban landscape in many cities is being decimated in much the same way as forests are being reduced to islands of plants.

Consequently, you have to utilise what space you have very effectively, as there just isn't enough room to grow many types of plants. For example, you could only have one avocado, walnut or pecan in the average backyard - and not much else!

For these types of situations, we need to better use vertical growing space, promote the use of dwarf and multi-graft fruit trees, and focus on growing herbs and vegetables for our culinary requirements. We may not be able to grow all of our food but we can supplement our needs.

Gardens

Many garden beds in urban settlements are narrow. The common bed shapes discussed in Chapter 2, such as circle, herb spiral, keyhole (mandala pattern) and plucking beds, are frequently used in suburban gardens.

Urban backyards might only contain zones one and two, and possibly some zone five. This means that you will be able to grow some vegetables, have a few fruit trees and maybe keep a couple of chickens. Essential to all gardens are compost (plant food), a variety of plants in garden areas and water. Here we discuss each of these in turn.

Compost

Compost is made when plant and animal materials break down into simpler substances. The process, which is normally slow, can be sped up by altering one or more of the necessary requirements for making compost. These requirements include air, high temperature, a good balance of plant and animal materials and water.

The balance of plant and animal materials is reflected by the carbon-nitrogen ratio, which should be about 20:1 for ideal compost making. You can build compost heaps from solely plant material, but the overall process takes longer.

Generally, the greater the amount of air available for the feeding micro-organisms and the less loss of heat (higher temperature), the faster the decomposition. So, by turning the heap or allowing more air to circulate through the pile, and by covering the heap so that heat is not lost to the environment, decomposition occurs quickly.

Many people find turning the compost heap hard work, so here is a simple solution. Recycle a metal 200 L (44 gallon) drum by cutting half of the lid away. Lay the drum on its side and start filling it

Figure 10.1 Use a 200 L (44 gal) drum as a compost turning device.

with compost materials. To increase aeration, roll the drum backwards and forwards or, if possible, use a small ramp as shown in Figure 10.1.

Many people just build a compost pile wherever they find some space in the backyard. Others have a set area. Your organic recycling centre can be built from materials such as old wooden pallets, or timber slats, as shown in Figure 10.2, or you may simply choose to place all of your organic wastes in the garden bed itself. A banana circle is an ideal compost heap, as food scraps and leaf litter are placed in the hollow in the middle and the surrounding bananas or other plants feed on the wastes

Figure 10.4 Small holes in bins such as these increase air flow through the compost pile. If the base is solid you should make holes in it to allow drainage of excess water.

Wooden Slat Compost Box

Figure 10.2 Compost bays are easy to construct.

as they quickly break down. You may have to check to make sure your local council allows you to have a compost pile in your backyard.

Some of the commercial plastic compost bins that you can buy (or are supplied by some local authorities) do not work effectively because air is restricted from reaching all parts of the pile.

Cutting holes all over the bin, including the lid, will increase air flow through the compost.

To minimise insects and other pests from entering the bin, glue some flywire screen over each hole. Holes in the lid are also important to allow rain water to enter the bin to keep the compost moist.

Garden areas

Much of the urban yard should be mulched. Sheet-mulched garden beds are common, and all other trees, such as fruit, chicken fodder and so on, are mulched, at least out to their drip line. Mulch increases the microbial activity of the soil, and plants growing nearby extend their root mass into the mulch, thus increasing their water and nutrient intake.

A banana circle creates a warm, humic microclimate.

Figure 10.3 A banana circle is a compost heap in disguise.

Mulch is under-used in the community, mainly because people are unaware of its benefits. The many conventional landscape practices that are used in gardens are cosmetic, and they are used to make the garden areas look attractive.

Permaculture, however, is about making these areas useful - and beautiful as well. So, we plant and maintain a variety of herbs, vegetables and other plants that contribute to our daily needs.

Permaculture is also about saving energy, so it makes good sense to place the food production areas near the consumers, whether these are humans, chickens or goats. You need to grow vegetables where you can see them - close to the house, along the paths to the chicken pen, or in view from the kitchen window.

When designing your garden beds, choose the plants most appropriate to your soil type and microclimate areas. If you have lots of shade (and you can't change this), plant shrubs that are shade tolerant, such as comfrey and angelica.

You can't expect to grow all types of plants successfully. Some like well-drained soils, while others tolerate clay and waterlogging. For example, walnuts are susceptible to root rot in heavy soil, but grow well in sandy soils.

Generally, most fruit trees are suitable for backyards. In particular, lemon, mandarin, orange, mulberry, apple, guava, banana (unless there is regular frost) and early varieties of plum survive in most climate types and are appropriate in small backyards.

Some types of fruit trees, including pears, peaches and nectarines, are often not as hardy, easily succumb to fungal diseases or are attacked by pests. Use fruit trees that don't get ravaged by birds or fruit fly, or become covered in fungus. Early varieties of most trees are best as they are less susceptible to attack by diseases and pests.

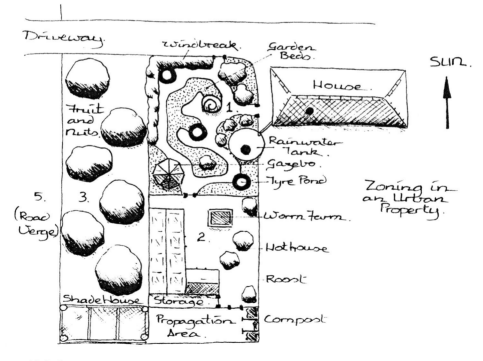

Figure 10.5 An example of zoning in an urban property. Food production areas are placed close to the house.

Fruit trees and garden beds need protection from damaging winds, even in urban backyards. Fast-growing, vertical windbreak shrubs and trees, such as clumping bamboos and banna grass, are ideal windbreak species. Bamboos are very useful plants and can be used for garden stakes, fence materials and thatching, and poles for pergolas and walkways (besides eating them, of course). How windbreaks function and how they are constructed is discussed in Chapter 11.

Paths should be narrow and can be made of any type of material, including sawdust, mulch, concrete and carpet. For slippery surfaces and slopes or even wooden steps, consider stretching and securing poultry wire across them. Your shoes will have better grip on the new surface. Steps can also be made from car tyres, filled with soil, as shown in Figure 10.9 on the next page.

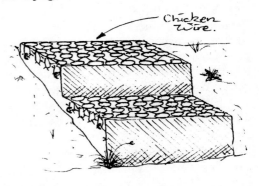

Figure 10.8 Stretch chicken wire over slippery slopes, paths and steps.

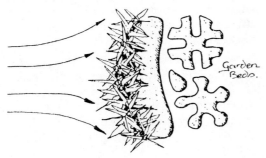

Figure 10.6 Windbreaks are also needed in urban properties to protect delicate plants.

Designers also need to consider garden bed access. Access to garden areas can be achieved by a variety of methods. For example, the simplest is probably stepping stones - of timber, slabs, stones or concrete. Paths can also be "living" by planting herb lawns, such as chamomile, lippia, *Dichondra* and thyme, in-between the stepping stones.

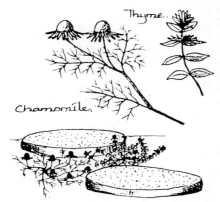

Figure 10.7 Stepping stones and herb lawns make useful, functional paths.

Ponds

Water is the most important element in permaculture design. Ponds and aquaculture were covered in Chapter two, so here we are just briefly looking at water in the urban landscape.

Even small ponds can be productive, and a large range of edible food crops can be successfully grown in water, such as taro, watercress, water chestnuts and Chinese water spinach (kang kong).

A word of caution: many parasites and pathogens (disease-causing organisms) have part of their life cycle in water, so make sure that you obtain healthy, disease-free plant stock.

You might consider introducing tadpoles caught from nearby swamps or creeks, as these will contribute to pest control in the garden - after they turn into frogs.

Frogs are a good indicator of the quality of a waterway. If the water becomes polluted or stale (low oxygen, anaerobic) the

Figure 10.9 Use tyres filled with soil as steps.

frogs will disappear. They might also disappear after predation by cats, so watch this.

An alternative to ponds is the bog garden. Here, a wet area can be developed, as many of the plants that live in ponds will thrive in marshy, wet soil as well.

The bog garden can be built by lining a hollow with plastic sheeting or clay and filling it with soil that is kept moist at all times.

The idea of using a bog garden, instead of a pond, in a school design is also briefly mentioned in Chapter 12.

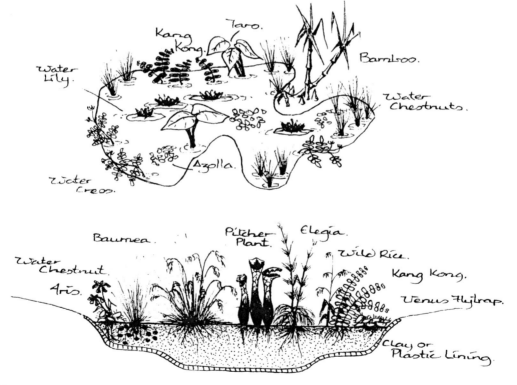

Figure 10.10 Top: Even small ponds can be productive. Bottom: A bog garden can replace a pond in a garden area.

Animals

Many small animals can be kept in the urban yard. However, the three most common useful animals are earthworms, bees and chickens. The use of animals in permaculture systems is further discussed in the next chapter.

Earthworms

Earthworms are the number one animal to have in the urban landscape. They are easy to keep and cultivate. Earthworms can be bred in a box, but they should be living in the garden! Here, earthworms do what they do best - improve the soil. You don't need an earthworm box to put your kitchen scraps in. Just dig the scraps into any garden bed and not only will they be turned into nutrients for your plants, the soil will benefit too. Garden beds should become your earthworm farm.

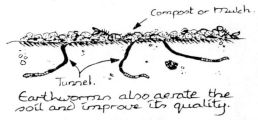

Figure 10.11 Earthworms are nature's cultivators.

There are many different types of earthworms, and the ones that are mostly used are the manure or compost worms, often small, red worms which are distinct from the larger earthworm found in some gardens and pastures. It is the manure worms which we are generally talking about, even though we just call them all earthworms, as these can breed quickly and be used to digest household wastes.

People who do not have gardens, or who rent a property, or may not have much room, can keep earthworms almost anywhere, even in a small box similar to that shown in Figure 10.12. Kitchen scraps, torn newspaper and other organic material can be placed in the box each day. An earthworm farm such as this might contain several thousand earthworms, each consuming up to their own weight in food every day.

An active earthworm farm of manure worms such as red wrigglers and tiger worms, properly managed, can digest up to a kilogram of food scraps each day. Provided that you keep them warm, moist, aerated and well-fed, earthworms will be tireless workers for you day in and day out. An earthworm farm can replace the need for a separate compost pile.

Although earthworms eat anything organic, they don't like too much of one thing, especially acid foodstuffs, so excess citrus peel, onion skins and tea bags can be placed directly into the compost heap. Earthworms also cannot digest bones, so

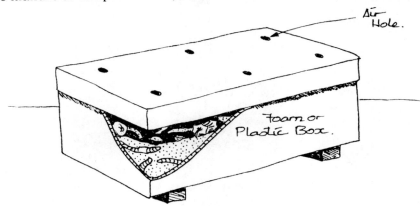

Figure 10.12 Earthworms can be kept and cultivated in a plastic, cardboard or foam box. Make sure that you have air holes in the top and bottom of the box.

after they have consumed the skin or feathers and soft tissue, put the skeleton, after a little crushing if possible, into the compost pile as well, or burn the bones and apply the ash to your garden beds.

Bees

Bees have a role to play in almost all natural ecosystems. They pollinate many different types of plants and therefore are essential in the plant's life cycle.

Not everyone likes to keep bees. Some people are allergic to bee stings and others have a fear about bee attacks. Some people can't keep bees because of local authority regulations - check with your local government before you buy and set up a hive. In some states or countries you may need a license to keep bees. Make sure you tell your neighbours about your plans too.

Bees are very productive animals. You can harvest about 50 kg of honey from one hive (double super) each year - more than enough for your own needs and to sell or give away to friends, family and neighbours.

You need to obtain a hive two boxes (supers) high. The bottom box is where the queen resides, and there will be a mixture of young larvae, pupae, pollen and honey amongst the honeycomb. You don't take any honey from this box.

The top box is where the worker bees store their honey. You can often extract this honey three or four times a year, depending on the strength of the hive and the season. For example, you don't usually harvest honey during winter and the bees are most active during the summer, so more honey is available then.

Bees like to live in a hive with the sun shining on it most of the time. This may be a problem in a shaded backyard.

The solution is to place the hive on top of a shed or garage roof. This will also mean that the bees' flight path will be high and this will reduce encounters with bees as

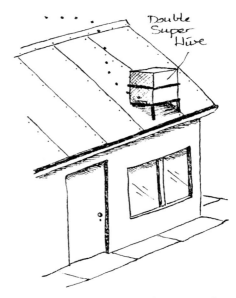

Figure 10.13 A bee hive can be mounted on a roof. This would only be practical if nearby trees provide some shade during the hotter months.

they fly to and from the hive during forage times.

If you want to keep the hive on the ground, you can still force the bees to take a high flight path as shown in Figure 10.14. Here a tree screen, shed wall or fence is used to make the bees fly high to get over the obstruction. Once over, the bees most often stay on the same flight path until they arrive at their destination.

If you've never kept bees or don't have the

Figure 10.14 Use a hedge or wall to force bees to use a high flight path.

necessary equipment, you might like to contact people from the Department of Agriculture, local authority or council, local Apiarist Association and bee product suppliers.

You may choose to let someone else manage your hives and extract the honey. The person who takes the time and makes the effort to extract the honey gets some and you get some. It's one of those win-win situations.

Setting up with a smoker, suit, hat and veil, hand tools, extractor, buckets and filters is expensive. For example, a hand-operated extractor might cost several hundred dollars, while a motorised one could be well over a thousand dollars (£500).

However, once you have tasted pure honey from your own hive, you'll never buy commercial blends again. You might consider sharing the costs for the necessary equipment with neighbours, or friends in different areas, who may extract honey at different times during the year.

Poultry

A couple of chickens are useful additions to any permaculture system. Again, there may be regulations about whether you can keep poultry in your locality or how many you can have - especially roosters! Check with your local authority. Three healthy chickens will provide one to two eggs a day for most of the year.

Chickens, ducks and geese can be kept in an urban backyard, but geese tend to be noisy at times and they prefer the larger open areas of rural properties. Ducks need to have water, at least a good bucketful so that they can dunk their heads and beaks. They also are noisy when you have a few of them.

Their water needs to be changed regularly as they will foul and muddy it quickly. While ducks are good at eating slugs, snails and most food scraps, they don't seem to break all things down like chickens do.

It is best to keep ducks, geese and chickens in separate pens, as geese will harass ducks and chickens, while ducks can attack chickens. Both geese and ducks will foul drinking water quickly, which may cause health problems for other poultry. Generally, ducks and geese suffer fewer diseases than chickens. You need to recognise that each poultry species has particular needs, and fulfilling those needs can best occur when they are separated.

In the urban situation most people just have chooks. There is a large range of different breeds of chickens and you can buy them from a variety of sources, including your permaculture friends, poultry breeders and backyard enthusiasts. Some are meat birds, some produce an abundance of eggs, some are black-skinned and some are small.

The smaller bantams are easy to keep and feed, and make good mothers, although their eggs are about half the size of the larger breeds of birds.

To recycle all domestic organic waste you only need three elements - a compost heap, chickens and earthworms. Give all of the household scraps to the chooks. Let them take what they want. After a few days, rake up what is left and place it in the earthworm farm or compost heap.

Of course, you will need to supplement the household scraps with other foodstuffs such as grains (cereals), herbs and greens to provide a balanced diet for the poultry.

All poultry make good pets. Children like to pick them up and hold and pat them. However, if you intend to eat any excess poultry that you breed, don't give them names. It is very hard to kill and eat Alice, Fred or Muffin.

If you have a big backyard you might consider a rotating chicken pen as shown in Figure 10.15. In a rotating poultry pen, chickens, ducks or geese can be kept in pen 1, while pens 2 and 3 are cropped with vegetables or planted with poultry fodder species.

When appropriate, the poultry can then be moved into the next pen and allowed to clean up the remaining vegetables or eat the fodder plants.

This is a useful strategy for the summer and autumn months when there is not a lot of green feed.

Poultry will weed, manure and aerate the soil, preparing the ground ready for the next crop. It is important that poultry have access to the housing shed, no matter what pen they are using. Access holes to each of the other pens are covered up.

Having three compartments or pens also allows one pen to be rested, one pen for the birds and the other planted with food for you or the poultry, or both.

For smaller backyards, either a chicken tractor or a deep litter system is ideal. In the deep litter pen, poultry are kept in a pen or shed and a thick layer of sawdust, leaves or even newspapers is placed over the floor.

Chicken manure and food scraps provide the necessary nitrogen for bacteria and fungi to break down the sawdust, papers and leaves, making a good compost within a few weeks.

All you do is rake out the layer once a month or so and put it on the garden, or if it does need a little more time to break down, throw it on the compost heap.

This method of placing leaves, paper or cardboard on the ground works well in the rotating chook pen too.

It is an easy way to make compost, so you don't need a special compost area or heap if you use this system.

A chicken tractor can also be used in small backyards, where it can be easily moved to different areas. It is essentially a portable poultry pen, often on wheels, so it can be easily pushed or pulled around.

Placing a mother hen and baby chicks in a chicken tractor is a good strategy to protect the young from predators like cats, foxes and large birds.

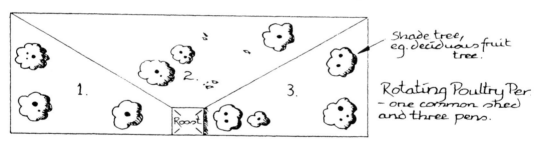

Figure 10.15 A rotating poultry pen.

Figure 10.16 A deep litter chicken pen.

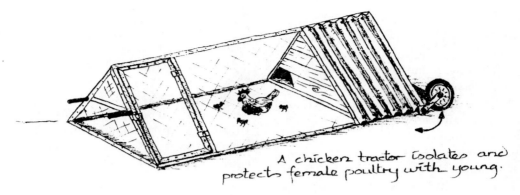

Figure 10.17 A chicken tractor.

Limited spaces

Permaculture is not restricted to the wide open spaces of large properties. You can practise the principles of permaculture and demonstrate the design concepts by building mini gardens and growing food on balconies, window sills and in hanging baskets, and even the occasional aquatic plants in an aquarium inside your home. You are only limited by your imagination and your design skills.

Balconies and windows

Many people don't have the luxury of a back and front yard. Although they may live in flats and units, they can still grow food in the space they do have available, and be clever about how they do it.

Many housing units have small balconies which, in most cases, are under-utilised for growing food. It is easy to make or buy stepped pot stands, free-standing mini-garden beds, or garden containers which you can screw or secure to a wall.

Herbs and plucking (green salad) vegetables can be grown in these stands so that

Figure 10.19 Free-standing mini gardens can be easily moved in a small area.

you have some food all year round. You can even utilise the underside of the balcony above you by growing plants in hanging baskets or securing trellis from one balcony to the next.

Windows and window sills in a house or shed can be used if there is limited garden space.

Shelves can easily be made to fit within the window frame, and window boxes and lattice or trellis can be secured to the wall either side of the window.

Again, a variety of culinary herbs and vegetables can be grown, including climbers such as peas, beans, nasturtiums and even passionfruit.

Figure 10.18 Stepped pot stands allow sunlight access for all plants.

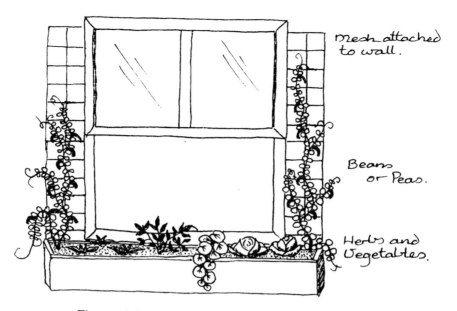

Figure 10.20 A planter box outside a window.

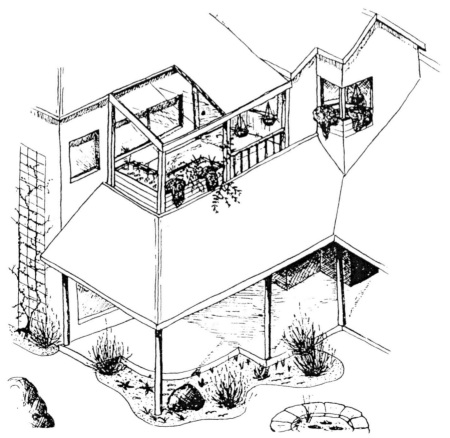

Figure 10.21 Use every available growing space in small areas.

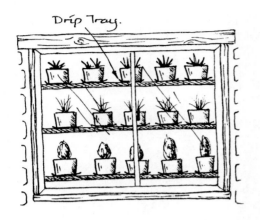

Figure 10.22 Build shelves inside a window frame.

Small backyards

There are many design strategies for small areas. These include having dwarf fruit trees in pots which can be rolled away when needed, trellis (both horizontal and vertical), and hanging baskets which can be secured to pergolas, existing large trees or verandah areas.

Trellis can take a variety of forms and be made from a variety of materials, including wire mesh, poultry wire, bamboo sticks and roofing battens. Ingenuity and common sense are used as we realise that in our yards we don't have junk, only treasured resources.

You can use some materials to construct a pergola-type walkway so that vines can be grown up and over the trellis, or you can place trellis vertically to grow climbers and vines.

Multi-graft fruit trees are another option. Multi-grafts have two or more different varieties of the same type of fruit on the

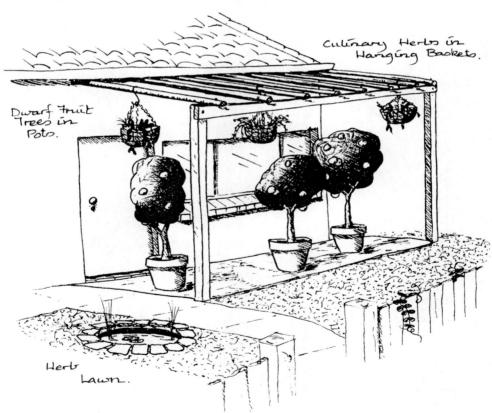

Figure 10.23 Dwarf fruit trees in pots are easy to move about.

Figure 10.24 Trellis and other structures use vertical height for growing food crops.

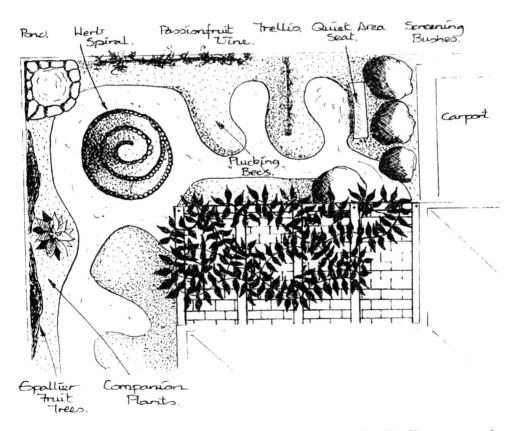

Figure 10.25 Fruit trees can be espaliered along a wall or vertical trellis. Vine crops, such as passionfruit, can be grown along a fence.

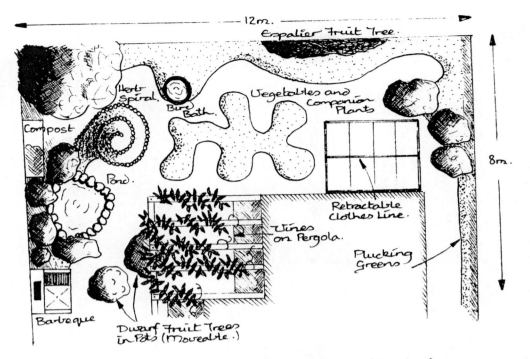

Figure 10.26 Garden beds need to be compact and functional.

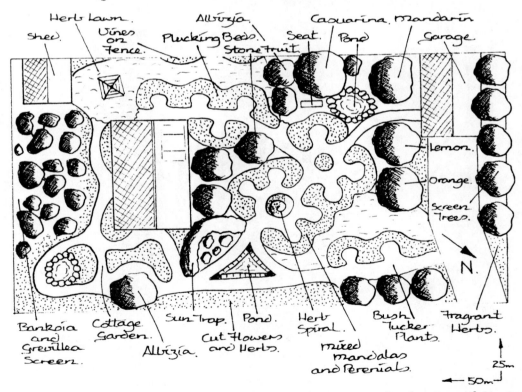

Figure 10.27 Permaculture design is an integrated, holistic approach to human settlement.

one tree. Apples and citrus can easily be grafted onto root stock, or an existing variety of tree, so you could potentially have fruit for a large part of the year from the "one" apple or lemon tree in the yard.

Before you graft different varieties onto your tree, or get someone else to do this, find out the fruiting times and other qualities of the varieties you are going to use. Otherwise, you may have one part of the tree growing faster and taking over the slower growing varieties, or you may have an abundance of fruit at the same time rather than the total yield spread over many months.

The simplest way is to place the seeds in a glass jar. Cover them with water and leave overnight, or for most of a day. Use a cheesecloth, stocking or clean kitchen-wiper cloth to cover the mouth of the jar - in place of the lid.

An elastic band or string is placed, or tied, over the stretched cloth to hold it in place. Place the seed jar in a warm place, such as on a window sill.

Discard broken and shrivelled seeds or seeds that are floating when you inspect the jar the next day. Drain the water off and rinse the seeds each day until the sprouts are large enough to eat.

Invert the jar after rinsing so excess water drains away. This daily washing of the seeds may take a week or more, depending on the actual type of seed used, the temperature and other climatic conditions.

Figure 10.28 Hanging baskets are ideal food growing gardens.

Germinating seed

A large number of seed sprouts are edible and can provide high levels of nutrition in the form of proteins, vitamins and minerals, and occasionally carbohydrates. Seeds of wheat, barley, sweet corn and lentils, along with the commonly sprouted mung beans, soybeans and alfalfa, are easily germinated.

Figure 10.29 Many seeds are easy to germinate in the kitchen.

Don't buy commercial seed packets, as some seed companies use fungicides and other chemicals to protect their seeds from diseases. Use only organically grown seeds or, better still, ones you have harvested yourself from your garden.

Seeds need to be from the last season's crop. Many seeds lose their viability after a year or more and so will not sprout, and those that do may contain little nutrition. Some sprouts mould quickly, so by trial and error you will know when to eat them. For example, lentils, peas and sunflower sprouts are eaten when they are less than two centimetres long, whereas mung beans and alfalfa are eaten when they are twice this length. Generally, the longer the sprout the more nourishment it contains, but if they get too long they do not taste as nice.

To germinate larger amounts of seeds you should use a cold frame or hothouse. Here, a glass lid covers a wooden or steel box so that winter sunlight can penetrate and warm the interior. Seedling trays are placed inside the box and watered each day. As seeds germinate, the tray can be removed and the seedlings pricked out into pots or placed directly into garden beds, provided that the danger of frost is over and the ground temperature is well above zero.

Figure 10.30 You can use a cold frame to germinate trays of seeds for the garden. All you need is a box with a clear glass or plastic lid. Heat is trapped and held inside the box.

My notes

Things I need to find out

11 Designing for rural properties

Whole farm plans

Planning is required for any sustainable farming development. The development of whole farm plans and the integration of all components on the farm further illustrates the importance of functional design.

The design of shelterbelts, fodder lock-up areas, water harvesting drains and woodlot areas must contain information about the maintenance and management of the system.

Economical analysis of the monetary outlay and income of farming and land management strategies has shown that putting in tree belts, building dams and undertaking earthworks for drainage and so on, can all have a high benefit-to-cost ratio. Essentially, this means that these strategies can make more money than they cost the farmer.

Not all benefits are financial in nature. Landcare, conservation and land management strategies will help improve the quality of the environment and also make the farm look pleasing - aesthetics are important to humans. Improving the quality of waterways, conserving and restoring natural wetlands, wildlife habitats and bush areas, and increasing the recreational value of land all benefit the community.

Farmers and land owners must consider what they will do to ensure the best return for money invested for land improvement work. When money is not freely available for all farm planning strategies, then some prioritising of all of the options should occur. Setting priorities helps the land owner to work towards realistic goals and the slow, ongoing implementation of the design.

An integrated whole farm plan allows all appropriate conservation and development practices to be blended together into a single system. The outcome will be good land management practice, as waterlogged areas dry out, dry areas get water, salinity problems are addressed and soil quality improves.

It might also mean that waterlogged areas are better used by planting wetland plants, and dryland species are used to maximise the potential of dry areas. Whole farm plans often have the following characteristics:

- paddocks are divided into homogeneous land units by soil type, natural topographical features, vegetation types, or water drainage areas.
- each land unit is managed and developed after examination of the potential stock carrying capacity, crop yield, amount of degradation and erosion, and soil fertility.
- the impact of developments and improvements such as roads, revegetation, water harvesting strategies and crop and pasture changes are considered in the light of the whole farm - in an integrated, holistic way.
- farm management strategies must incorporate sustainable practices, such as maximising nutrient recycling, minimising energy and resource use, maintaining land productivity, preserving natural ecosystems and increasing species diversity, as well as making sure that the farm is profitable.

The successful implementation of a whole farm plan is a long-term process. It literally takes many years before financial benefits to crops, pastures and animals result from tree planting and other sustainable land practices.

However, even though land owners make substantial capital investments on improvements to the soil, fences, fodder areas and water catchments, the value of the farm land increases and other benefits to the environment follow.

Most farmers view replanting and revegetation initiatives in terms of economic return. They need to be convinced that shelterbelts, regeneration of native forest areas and replanting of riparian (stream or river bank) areas will add value to their property, make the use of the land more sustainable and give a potential income in the long term.

Income from woodlots and specialised timber may take ten years or more. Even fodder tree species have to be fenced from stock for two to three years before cutting or grazing is allowed. Farmers should be encouraged to become foresters and to replant trees in degraded areas and near remnant vegetation patches. Furthermore, conservation must be seen as good business.

Economic viability must be equated to ecological stability. For example, on small and broad-acre properties the practice of minimum tillage should be investigated.

Nature doesn't plough and turn the soil, and nor should we. You may need to break up compacted ground, either by using a chisel plough or, in some cases, ripping to plant tree belts.

Ripping along the contour also increases water absorption by the soil. This can be an effective strategy as this technique is cheaper than building swales.

Windbreaks and shelterbelts

Severe winds reduce food production. The most damaging winds are hot and dry or those coming from offshore laden with salt. These types of winds increase the transpiration and desiccation of the plant. Cold winds during winter will also adversely affect many subtropical species.

Strong winds can do physical or mechanical damage and the ferocity of severe winds needs to be tempered. Many food producing trees such as kiwifruit, macadamia nut and bananas are easily damaged by wind, with food yield reduced.

Windbreaks should be positioned to counter the effects of severe winds, whether they are prevailing or not. Prevailing winds are those blowing most often from a particular direction.

In some places your windbreak tree belt may have to be placed on the eastern side of the house, in other areas the western side. In other places, the prevailing winds could be south-westerly or north-easterly. The wind direction of prevailing winds in winter will differ from those in summer, and cooling, summer afternoon (ocean) breezes are often in opposite directions to night-time (land) breezes.

You might like to re-read Chapter 3, which deals with sector planning, to have a better understanding of where windbreaks are placed and how the energies that move through a system are directed and controlled.

Figure 11.1 Tree shape gives some indication of the direction of prevailing winds.

Land and sea breezes occur because of the uneven heating and cooling rates of land (soil) and water. During the day, the land heats up faster than the sea. Hot air rises above the land and is replaced by a cool sea breeze which usually blows in the afternoon. At night, the land cools faster than the sea and cold air from the land flows towards the ocean.

Sometimes, wind blowing across the land heats up and it is not uncommon to experience hot, dry land breezes early in the morning. Water tends to heat up and cool down slower than the land.

There are several terms which have the same meaning and hence are interchangeable. For example, shelterbelts are essentially windbreaks. However, we sometimes use shelterbelts to describe rows of trees surrounding agricultural land while windbreaks may refer to those trees protecting buildings, orchards and garden areas. Shelterbelts and windbreaks are planted in a direction which is about 90° (right angles) to the prevailing winds. Shelterbelts, as the name implies, are often arranged so that stock can find shelter and refuge in them, especially when giving birth and during storms.

Windbreaks protect the soil, preventing it from drying out and minimising the loss of topsoil by wind erosion. Greater production occurs in areas protected by windbreaks. Trees make the best windbreaks. Solid fences should never be built as they can be pushed over by the force of strong winds and they cause turbulence.

As a general rule of thumb the area protected by a tree windbreak system is about twenty times the height of the trees. In Figure 11.5, if the trees were 5 metres high, the distance that is protected on the leeward side is about 100 metres.

Figure 11.2 Land and sea breezes.

Windbreaks should be a minimum of three rows, but preferably five rows in some circumstances. Five row windbreaks can be used for paddock fence lines within the property. These give protection to the

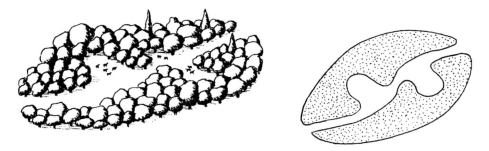

Figure 11.3 Shelterbelts are windbreaks that provide protection of stock.

paddocks on both sides of the windbreak system because wind speed is reduced and winds are ramped over them.

Trees also reduce the amount of wind passing through the area. At least half of the wind may still pass through, but at a much slower, less damaging speed. Trees further reduce wind erosion by trapping sand and silt.

Large trees which are slow-growing, such as oaks and carobs, should be planted in the middle rows (of a five-row set) and the fast-growing (and nitrogen-fixing) trees, such as wattles, tagasaste and casuarina,

Figure 11.4 Windbreaks are used to deflect wind upwards.

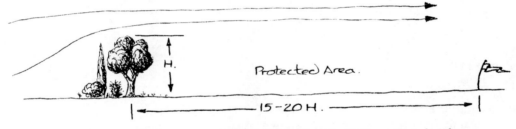

Figure 11.5 Tree windbreaks offer protection for 15 - 20 times their height.

Figure 11.6 Poorly designed windbreaks can intensify the problem. Shrubs and trees should have foliage to ground level.

are planted on the outside. These offer protection while the more productive and often commercially-viable species are growing.

Like all permaculture elements, windbreaks should also serve several other functions. For example, windbreak trees can also be used for bee forage, sources of mulch and firewood, screens for noise, absorbers of pollution, shelterbelts for stock and suntraps for climate modification. Species which can serve as fodder and windbreak include acacia, tagasaste, carob, oaks and leucaena.

Well-designed windbreaks can also slow a fire to one-tenth of its speed. Windbreak trees, which are used as natural firebreaks as well, should have high moisture contents, low levels of flammable resin and oils, and should not shed leaves and branches.

Whenever any planting occurs, the sequence of shrubs and trees should follow nature's pathway. Pioneer trees need to be planted first on denuded and degraded land.

Once these are established the larger climax species can be planted. Remember that the pioneer species such as acacias, casuarinas (or holly and brambles in colder climates) are good windbreak species and protect the climax species such as oaks, nuts, beech or carobs.

Stock need shelter from cold winds as

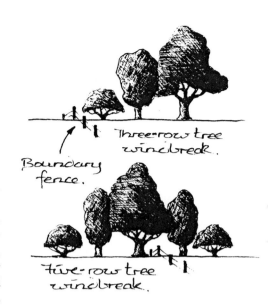

Figure 11.7 Windbreaks should contain several rows of trees.

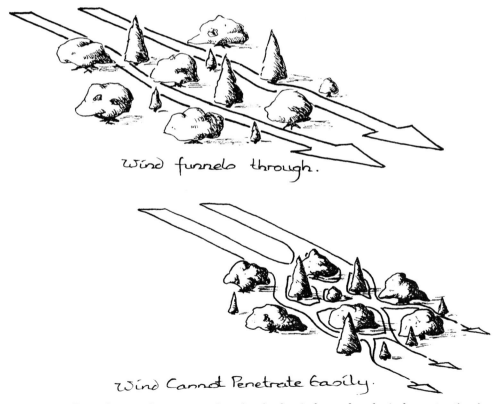

Figure 11.8 Trees have to be staggered so that both wind speed and wind penetration is reduced.

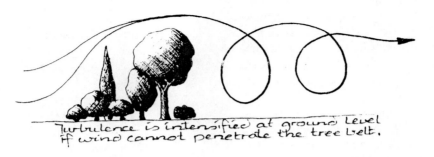

Figure 11.9 Wind must be able to penetrate some of the tree belt, or turbulence will occur.

many will die in extreme conditions lasting several days. Shelterbelts of trees provide refuge as well as fodder during these lean times.

Stock production decreases both during extreme cold and extreme heat. In fact, shelter is crucial to stock productivity, as open pastures will decrease milk, wool and meat production. The placement of these belts of trees, fences for stock yards and water supplies are crucial to any property containing stock.

Shelterbelts can also be productive in other ways - as sources of firewood, honey, mulch, building timber and edible crops. For example, chestnuts for humans and

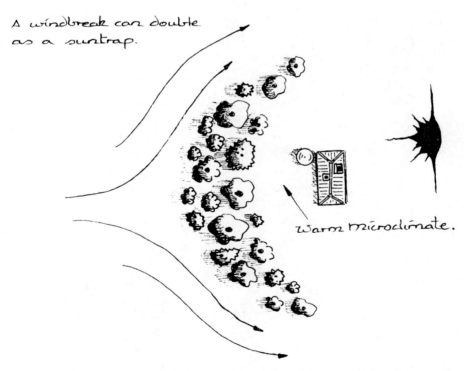

Figure 11.10 Windbreaks should be curved so that wind is deflected as well as slowed. Banna grass (*Pennisetum* spp.) or clumping bamboos are useful windbreaks. Here a windbreak also acts as a suntrap.

Figure 11.11 Nurse trees such as tagasaste or wattles are used to protect wind- and sun-sensitive plants such as macadamia nut trees. Nurse trees can be continually pruned as the protected tree grows.

medics. Finally, larger nut and berry trees, such as oaks, leuceana and carobs, should be used as they drop pods or nuts with high carbohydrate and/or protein content during the lean autumn or fall periods.

Alley cropping practices in the paddock should follow the contour as much as possible (again to harvest water to supplement their needs). Even small pockets of trees would effectively improve pastures and reduce erosion. For some windbreaks on a slope, place swales every 10 m or more to capture water to provide additional watering of the trees. The higher slope areas tend to dry out first so you should rotate stock from high areas downwards.

Tree belts have the added advantages of: stock shelter and shade; microclimate changes so that crop areas are more productive; greater animal and bird life on the farm; source of nectar and pollen for bees; and source of firewood, construction poles, fencing posts and timber. All of this means, in turn, greater stock carrying capacity and greater opportunity to diversify farm income, by developing aquaculture in the dams, honey from bee hives and timber for sale.

On large agricultural areas, shelterbelts may be spaced about 20x the height of the tallest tree. This corresponds to the amount of protection that this shelterbelt offers.

Planting trees close together ensures canopy closure which, in turn, minimises wind infiltration. However, light is restricted from penetrating to the ground layer and few plants will grow as understorey. Understorey shrubs increase the ecological diversity and stability, and they are important in the tree belt system. Good design of shelterbelts will allow thinning and pruning without affecting their effectiveness against wind.

There is no magical number for the percentage of farmland which should be planted or replanted with trees. Even at

oaks for stock. Fast growing, potentially commercial timber species include paulownia (*Paulownia tomentosa*), blackwood (*Acacia melanoxylon*), silky oak (*Grevillea robusta*) and oaks (*Quercus* spp.) generally. Fast-growing nursery, stock feed and nitrogen-fixing trees include wattles (*Acacia* spp.), tagasaste (*Chamaecytisus palmensis*) and *Albizia lophantha*.

The arrangement of tree belts with arable land in-between, which is used to grow some sort of crop, is called alley cropping or hedgerow intercropping. Even though arable land is lost in the planting of these tree belts, the total production of the land increases and it is a worthy system to adopt.

Alley cropping, which is also known as agroforestry or forest farming, is growing trees, usually as tree belts, in conjunction with agricultural crops or pasture and domestic grazing animals, or both.

The other issue for permaculturalists is that we should be advocating the use of perennial pasture crops rather than annual pastures. Perennial pasture plants include herbs such as comfrey and dandelion, and fodder shrubs and trees such as *Coprosma*, willows, poplars, tagasaste and

15% tree cover, the loss of production of pasture is more than compensated by increased crop production and potential income from tree-based enterprises.

Furthermore, try to design zone 5 areas so that they are linked to natural bushland, remnant bush areas and wildlife corridors. Consider planting rare and endangered species from these areas - especially on steep slopes greater than 15 to 18° - as one way you can contribute to their survival.

The replanting of indigenous tree species in areas which were once forested but are now depleted, such as sparse bushland and agricultural areas, is called "reforestation". Afforestation is planting trees in areas which are devoid of trees, such as desert areas. Even so, many of these desert areas were once forested, but are no longer because of climatic change and human intervention.

To get plants established, some preparation and care should be undertaken. Direct seeding is a cheap method of tree planting, and provided the seeds have been pre-treated, such as pelletising the seed or scarifying the seed coat, success should follow.

Small seed can be mixed with sand before spreading. The land owner is advised to initially proceed on a small scale, until a working system is developed and established.

Overgrazing a weed-infested area by stock, before planting, may give new seedlings a greater chance of survival. Generally, seed planting is more successful than tubestock planting, as germinated seeds send down long roots in search of water, whereas roots are often pruned (air or physical) in tubestock pots and do not re-establish quickly in the ground.

Poor weed control is perhaps the greatest hindrance to plant establishment.

Animals in the system

Each farm animal has its own special requirements and its own special functions. Large animals such as horses and cows require large amounts of water and feed, are difficult to control and pen, are less efficient at converting plant material into animal protein and have a low reproductive potential.

Figure 11.12 Trees planted close together (1 m apart) tend to be tall and narrow while those of the same species planted further apart tend to become more bushy and widespread.

As you can imagine, most small animals, such as sheep, guinea pigs and poultry, are much the opposite.

The selection of the number and type of animals that are placed in the system depends on factors such as:

- climate and environment conditions. Some breeds of sheep, for example, prefer dry climates, while others survive on poor hill sites or in fertile lowland areas.
- size of property. For example, you may only have enough land to carry one cow but ten sheep.

- breeding habits. Some animals produce large litters (pigs) while others only one or two offspring (cows, sheep). You may need to assess the potential damage that could be caused by domestic stock which escape and become feral, including their effects on native animals and plant populations.
- forage and fodder requirements. Chickens require a minimum of about 200 g a day, while sheep consume 1 to 2 kg and cows usually greater than 5 to 6 kg.
- stocking rate. For example, only one cow for every ten acres, one sheep an acre and so on. This depends on the carrying capacity of the land.
- purpose. You might choose stock for meat, wool, milk or manure. You may want an animal to keep the weeds down in the paddock. If you want land cleared, and shrubs, trees and weeds removed, then get pigs.

Pigs can be contained by electric fencing, as shown in figure 11.13. A simple electric line is all you need. The pigs will "learn" quickly and test the line periodically - or push one of their mates into the electric fence to see if it is still on!

- personal preference. Some people prefer particular animals and not others. Since you will be looking after the animals, take time to consider your opinion.
- husbandry needs. Animals have particular needs. For example, sheep may need dipping and drenching (for lice and parasitic worms respectively), while cattle might have to be treated for mastitis, ringworm or warbles (lumps under the skin caused by the grubs of warble flies).

In planning for stock, make sure you consider both the land capability and the sex ratio. For example, one sheep per acre, one cow per ten acres, one ram for fifty ewes and one rooster for 10 hens. Stocking levels need to be low enough to allow plant regrowth. Some characteristics and requirements for particular animals are listed in the tables that follow.

Stock lock-up areas are a good way to provide feed all year round. In Figure 11.14 stock are maintained in pen A for a few weeks and then moved into pen B. This set-up and rotation can be repeated all along the farm boundary (if serving as a windbreak as well), or wherever stock are kept. Stock should only be locked up in the fodder pens for a short time, then moved on.

Figure 11.13 Pigs can be contained by electric fencing.

Animal	Size	Financial commitment	Feeding habit	Reproductive fecundity
Horse	large	large	grazer	low
Cow	large	large	grazer	low
Sheep	medium	small	grazer	low
Rabbits	small	small	grazer	high
Pigs	medium	medium	scratcher	high
Chickens	small	small	scratcher	high
Goat	medium	medium	browser	low

Table 11.1 Some characteristics of animals.

Animal	Uses, notes	Manure as fertiliser	Special requirements
Horse	draft animal	fair	shelter
Cow	milk producer, manure	poor, slow release	consistent good quality feed for milk
Sheep	wool or meat	fair	shearing
Rabbits	meat, fur	very good	shelter, protection from predators
Pigs	meat, omnivorous diet	poor	shelter
Chickens	eggs, meat	good	shelter, protection from predators
Goat	milk producer, need strong, secure fencing	good	consistent good quality feed for milk

Table 11.2 More characteristics of selected animals.

Note: 1. Animal manure for use as fertiliser is judged by the percentage of nitrogen and phosphorus.
2. Grazers mow the grasses to near ground level. Browsers prefer new leafy shoots and buds of shrubs and trees. They eat above the ground.

It is important to only allow the stock a short stay in each enclosure, as plants that are too badly eaten and pruned may never recover.

Different types of stock can be rotated through the same area, one after another. For example, cattle could be contained first. They will chop off higher limbs, some of which will fall on the ground or hang lower. After a week or two (depending on the number of plants and animals in the pen area) they can be moved on and replaced by sheep which will browse the broken and lower-hanging branches.

Stock must be rotated through grazing areas. Plants that are overgrazed react by producing and secreting toxins to deter browsers. Moving stock on also allows trees and shrubs to recover and produce new shoots.

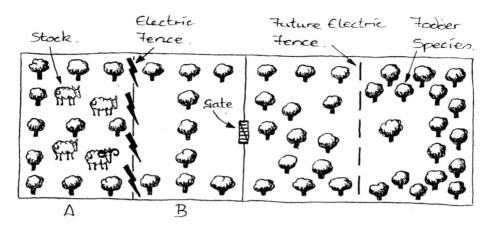

Figure 11.14 An example of a stock lock-up forage system.

A moveable electric fence is used in this intensive strip-grazing paddock. Fodder species such as wattles (e.g. *Acacia saligna, A. longifolia*), tagasaste (*Chaemocytisus palmensis*), albizia (*Albizia lophantha*), honey locust (*Gleditsia triacanthos*) and leucaena (*Leucaena leucocephala*) are grown, usually in rows about five metres apart so that machinery can be driven between them.

An alternative system combines fodder with a windbreak strip. Fodder trees (as mentioned above) and shrubs such as tree medic (*Medicago arborea*) are planted as a windbreak along the length of a paddock. Stock can be directed into the fodder strip, which is usually used as an occasional feeding lot, especially during the "autumn feed gap", or they can forage along the fence line at all times.

If electric or conventional fencing is too difficult or too costly, or other factors limit this type of system, you can simply cut plant material and feed the stock at your

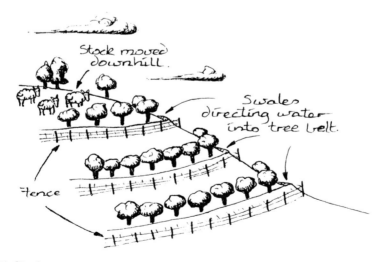

Figure 11.15 Stock start at the top and are directed downhill as the paddock dries up or tree damage is imminent.

will. Prune trees at waist to shoulder height. If you let the trees get too big, stock won't be able to reach and you'll need a ladder to prune them yourself.

Commercial slashers and cutters are now

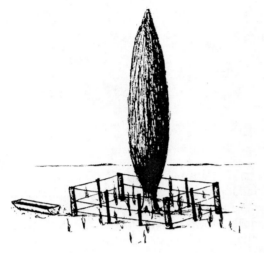

Figure 11.16 Suckering or running varieties of poplar, wattle and bamboo can be used as a fodder provision strategy for stock. As long as the parent tree is not devastated or eaten, the tree will survive even if the suckering growth parts are eaten to the ground. Established trees may not even need fencing.

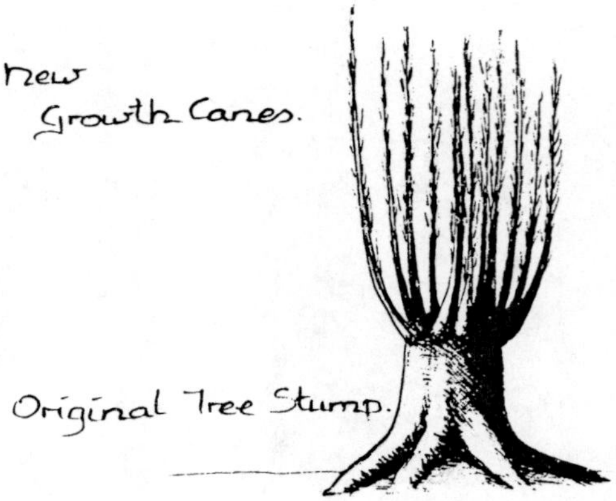

Figure 11.17 Coppicing of willow species produces many wands or canes which are used for craft work or stock fodder.

available, so trimming the plants occasionally by machine may be practical, if you can hire or find a contractor. Tree cutters are expensive to buy, but for large farms, having your own machine may be necessary.

Many trees can be coppiced. This means that you can cut the tree at stump height and it will re-grow. The cut foliage can be used as fodder, building timber or firewood. Poplars and willows are well-known coppicing trees.

My notes

Things I need to find out

12 Permaculture in schools

Schools are a special situation in permaculture design. No-one lives there, students grow up and leave each year, there is no permanent ownership of the garden area, no-one looks after it all of the time, and much of what we put in may be cosmetic and for demonstration.

However, what better place is there to become partners with nature and to learn about the water cycle, nutrient cycles, earthworms, food chains, soil and foods than a school garden? Teachers do not have to take classes long distances to observe nature. They can create gardens and bush or mini-forest areas right at the school.

Gardening allows teachers to instil a love for nature and for the land, and if students feel intimately involved with nature, their concern for environmental problems will be long-lasting. Some of the positive outcomes from building gardens are the sense of pride and accomplishment in meeting success, and feelings of self-worth as things start to grow and change occurs.

Even so, it is difficult to grow heaps of vegetables because other students, vandals and community members may raid the garden and destroy or steal things. Chickens can disappear after a weekend, ponds may be spiked and then leak, and plants are uprooted and thrown about the place. This doesn't happen all of the time or at all schools. But it does occasionally happen and teachers and designers need to be aware of this. In other words, schools have different needs that must be considered in the design process.

A needs analysis

In working with children and schools, as with any permaculture client, you need to determine:

- the needs and wants of the children, teachers and the school.
- what the budget is (how much can they spend).
- how much time and energy is available to implement, maintain and develop the garden area.
- what resources are available - both on site and in the local community.
- the potential site - its limitations, existing structures and positive qualities.

The biggest stumbling block for the development of school grounds is money. Usually the school administration and staff are on-side but lack of resources often dampens the enthusiasm. Materials can be obtained from a variety of sources, such as donations from the local nurseries and landscape suppliers, but often you have to beg students, parents and staff for plants, pond materials and other donations. Accessing resources and materials is an ongoing project, and this needs to be considered in the initial designing stages.

To determine the needs of a school, the budget and level of support in the community, you really need to discuss your ideas with a wide number of people, some of whom will include the teachers who want to use the area, school principal, other school staff, the gardener, and members of the school council and/or the parents and citizens (or parents and friends) association.

Determining resources

The resources available should be assessed. Does the school dump its lawn clippings or are they composted on site? Is there money for plants, mulch and compost? How much land is available? Is a source of water and irrigation near the proposed area? Ask yourself these types of questions.

Resources can take the form of people, materials, money and energy. Material resources can be assessed, for example, by getting students to conduct surveys which determine the amount of building supplies, newspapers, carpet or underfelt, tyres, sleepers and bricks which the school can obtain from the local community - either by way of donation or, at least, at reduced cost.

Securing "people resources" should be a high priority. Staff training and professional development are crucial to the success of a school garden area.

Individual teachers don't have time to spend maintaining the garden, so this responsibility should be shared.

It is best if all staff could have some training in permaculture, but this may not be possible. At the very least, the school should make a financial and supportive commitment to make sure several staff have an active interest in the garden.

Having a school gardener who is sympathetic to permaculture ideas is an added bonus for any school. However, schools which do not have a regular gardener, or which rely on community or parent support to maintain the school lawns and garden areas, should not be seen as disadvantaged.

There are many parents and community members, not directly associated with a particular school, who jump at a chance to put in gardens and gain practical experience after they have done a permaculture course. Some of these may be working towards their Permaculture Diploma and are more than willing to devote some regular time and energy to the school grounds.

Guidelines for designing school grounds

Involve children as much as you can. Teach them about design. Get them to measure the area and do scale drawings.

Show them how to build sheet mulch garden beds, visit established local permaculture properties and discuss the main principles and concepts of permaculture. Many students will begin to understand the ideas after they can visualise what they can do at their school. Students learn by seeing and then doing.

It may not be possible for the students to come up with a fully-functioning design, but unless they have some ownership, along with the teacher, of the plan, the future success of the garden area is not guaranteed. Children get excited and willing to work when they know that what they are doing is something they have contributed to. You may be surprised at the number of really good (and innovative) suggestions they make about the garden and site development. Lead them gently through the process.

Your role, as a designer, is to oversee the development and offer expertise and help. It should not be your job to draw the design without input from the "users". Keeping this in mind, here are some useful guidelines when considering building gardens in schools:

- consider the ages of the children. Small areas are great for small children, not so good for teenagers. Garden beds may have to be smaller than usual so that children can access all areas easily.

- children like their own garden beds. Groups of two or three work well together and can easily build, plant and look after their beds. Individuals should be allowed to develop their own garden if they ask, but encourage group co-operation. Remember that we are trying to teach life skills and community values, as well as gardening.

- sometimes different classes want a part of the garden. It may be better to consider building small garden areas nearby each classroom, rather than one larger area away from the school. Smaller

Figure 12.1 Small, easy-to-build and maintain garden beds should be developed by different groups of students.

gardens, solely the responsibility of one class and one teacher, are more intimate for the children. Alternatively, allocate different areas in the larger garden for those classes and teachers that do want their own space.

- will other groups of people, besides those directly building and maintaining the garden, be using the area? How can you use other teachers and their classes in the garden? Can you ask the art department to get their students to design and make clay-fired birdbaths? Do they want to display some of their sculptures in the garden? Will manual arts help build the seats you want to put in the area? Do any science teachers want to set up and stock a pond? If you involve everyone, then everyone will own the garden.
- will the produce from the garden be sold, taken home or given away? Is the school canteen willing to buy foodstuffs from the students? What kinds of foods would the canteen want? Gardening gives a direct connection to our food source. Many students initially fail to see the connection of what they are growing to the food they eat at home.
- some garden beds could contain plants for propagation work. These stock garden beds may be small beds of particular herbs that are used for cuttings and grafting work. Beds could have different themes. For example, one bed for culinary herbs, another for medicinal herbs and another for pest repellent herbs. All students could then take small pieces of these plants from the stock beds for their own use.
- outside activities should be organised and related to the curriculum. Many educational objectives and outcomes can be covered by simple, fun-to-do activities in your new outside classroom.
- other types of activities lend themselves to schools. For example, nesting boxes for birds, possums and bats could be studied by science students. Place one or two

Figure 12.2 Bird boxes and bird baths in the school grounds are useful teaching aids.

bird baths and/or feeding trays in the garden - you'll be surprised what you will attract. Build a weather station that holds equipment which measures daily temperature and barometer changes, humidity, rainfall and wind direction and speed.

- an energy audit of the school buildings would provide information about electricity consumption and waste in the classrooms. This could lead to the development of an energy management policy and a desire to reduce school operational costs.

At all times we are trying to teach young people about the six R's - reduce, refuse, reuse, recycle, repair and rethink.

- as schools move more into environmental education other school-directed activities follow. For example, the school may become the recycling centre for the local community.

Glass, paper, metals and plastic are brought in by students or their parents, maybe once a week. Money raised from this venture is invested back into the school to provide equipment and teaching resources.

- in some circumstances the school garden could become the community garden. Parents and community members could help students with garden development and/or be responsible for areas themselves. They may grow things to take home or for use by the school.

There is a general push by governments and education authorities to better use schools as this valuable resource is under-utilised at present.

Community gardens are one way to ensure a win-win situation. It also helps alleviate some of the potential vandal problems mentioned earlier.

One of the pleasing outcomes is that students start to teach their parents. Children soon build ponds and gardens at

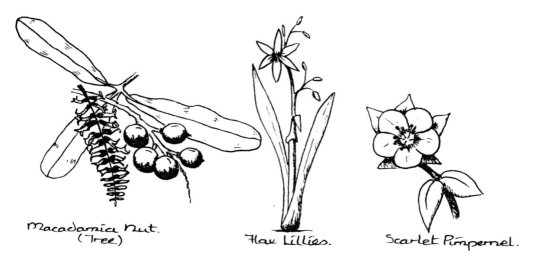

Figure 12.3 Bush tucker and bush medicine plants make useful additions to any garden.

home, plant vegetables and develop simple earthworm farms and compost heaps after they see how easily it is done at school. The skills that children learn help build their self-esteem as they realise that they have the ability to grow things and that they have a role to play as part of a co-operative team - a role that they learn at school which they continue to develop throughout their life as part of the local community.

Practical design considerations

The development of the school grounds should be within the parameters of the school plan or the vision for the future direction of the school. For example, if the school wants to conserve water and plant a native garden, then consider native bush tucker and bush medicine plants. Many schools already have a master grounds plan. A permaculture design of one particular area needs to complement any existing school plans.

If natural bushland or woodland is on the school site, develop a strategy to replant, maintain and protect this area. Part of your design for the school should address this issue. You may be able to develop a walk or nature trail, register the bushland as a heritage value area, or seek state or local government assistance to help preserve the area from overuse.

The number of activities, garden ideas and design strategies are endless. Teachers and students will be able to experiment and see what works best for them and their school environment and situation. However, here are a few ideas and practical considerations that have worked well in schools:

- where will tools such as shovels, pitch forks, hand trowels, gloves and wheelbarrows be stored? What about potting mix, plant pots, watering cans and seeds?

  Can you put a lockable shed in the design so that all of these are available on site? Worry about buying or obtaining one later on.

- plan for open space. You have to get at least one class in the garden area at any one time. This could be up to thirty students - where are you going to put them all?

  Consider bench seats and areas of

herb lawn where students can sit and work. Not everybody has to work on the garden beds during the lesson. Some will want to plan, record, draw and think.

- main pathways have to be wide. It is not practical to have the half-a-metre path that you have at home in the school garden area. Plan for a couple of students walking side-by-side or wide enough for a wheelbarrow and some leeway so accidents don't happen.

A pathway about 1.2 m wide is ample for school gardens. Smaller pathways are more suitable for individual (student) garden areas, or to lead to quiet places within the garden.

The typical size of a Mandala keyhole garden is not appropriate in some schools. You either have to make the paths and keyholes larger or consider making a double Mandala with keyhole beds on the outside as well.

- secret or quiet places are important. Children like to hide, especially from teachers, and secret garden areas provide sanctuary. Young children are quite happy to play hide and seek, older children hide for different reasons, so teachers should not make secret places too secret or overgrown.

It is possible to include areas for student privacy, yet still make these visible from the staffroom or classroom.

- seating is essential. Simple log or bench seats are more than satisfactory, but rock or carved seats can really look great.

Older students do like to sit and talk. Younger students are too busy running around.

- plan for some garden areas to be shaded. This could mean placing some of the seats under large trees or using structures like walkways, vines and pergolas.

In hot climates, shade from the sun is a critical factor in design, as students will only use and sit in the garden area if they feel comfortable.

Many schools have policies about

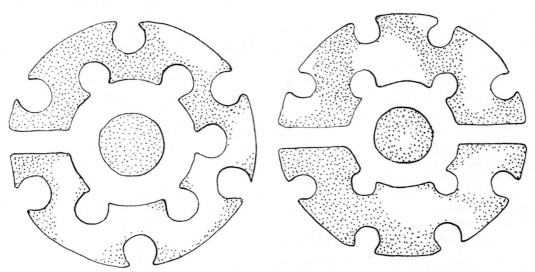

Figure 12.4 Two different ideas for a double Mandala which provides greater student access to the garden beds.

Figure 12.5 Students value a quiet place to sit and talk.

- outside activities and the precautions needed to be taken by students, including wearing hats and sunscreen cream, and the length of exposure time in the sun. There are great opportunities to design areas for shade so that students can safely work in the garden.
- developing your own shrub and tree nursery will help in reducing costs of obtaining plant stock for the garden. This can be as simple as a potting area, small shadehouse (covered frame) or a series of benches in a hothouse for plant propagation.
- compost areas may cause some minor problems. Freshly-made compost tends to smell, especially when you use animal manures and

Figure 12.6 A simple nursery can be set up to raise seedlings for your garden beds.

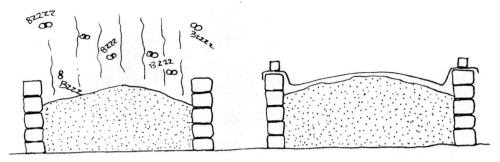

Figure 12.7 Compost may smell if it is not covered up.

it is summer! If there is a likelihood that odours may occur near classrooms, consider shifting the compost area to another part of the school away from buildings. This may not sound the best strategy from a permaculture viewpoint, but you have to consider the needs and comfort of teachers and students.

Open, uncovered compost and manure piles could attract flies and pests. It is worthwhile to build proper storage bays from wooden sleepers or concrete slabs so that these can be easily covered with hessian, carpet or underfelt to reduce the smell and pest problem while the compost is being made. While the end product may have a "nice earthy smell" the process of decomposition produces a range of noxious gases if the ratio of plant and animal materials, water and oxygen are not right.

- there are often some restrictions on building ponds in school grounds. Some education departments or authorities have rules and regulations, mainly for safety reasons, about water areas. Find out what these are. Your pond may have to have a weld mesh cover over it, or only a certain depth or a certain size.

Schools should be fun! Developing an outside classroom, where students can both learn and enjoy, is a sensible way in which educators can effectively teach the "curriculum".

Some students only seem to learn by working outside in the garden or doing practical activities. Others thrive in the more academic sphere. All students, however, need stimulating lessons to hold their interest. What better way is there than to go outside and observe and learn from nature?

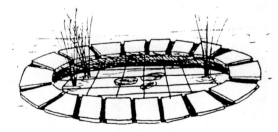

Figure 12.8 Ponds may need to be covered to protect young children. Alternatively, consider a bog garden which will not need the mesh covering.

My notes

Things I need to find out

13 Permaculture communities

Hamlet - to be or not to be?

Being isolated in the urban sprawl may not be the best way to live. Some people feel threatened by the urban environment and they will spend lots of money on security. Suburbanites are renowned for staying in their homes and not interacting with the people around them.

The type of housing developments most of us live in do not encourage neighbourhood interaction. Some urban developments are successful, such as the general move toward mixed-density "green street" areas.

Even so, many people don't even know their neighbours' names. There is an alternative - community living.

The structure and beliefs and values of communities vary enormously, and one that suits your lifestyle and needs can usually be found. If not, consider starting your own group.

Village or community living has the potential to reduce the general cost of living for individuals and families. There could be some relief from the economic burden that many people carry with the usual mortgage on a house in the suburbs.

The dream of owning your own home is really just that, a dream. Money lenders, banks and other financial institutions own the security of almost everyone's home.

If you want a different lifestyle, it may be achieved more easily if you are debt-free. Village living may give you the opportunity for this to happen, although this does depend on many other factors such as your financial position, because you still may need to borrow money.

It wasn't that long ago that people lived in small villages dotted about the countryside. These little hamlets, as they are also known, amalgamated as houses were built between them on main roads. As the population increased, the "village" sometimes grew to become regional towns and finally larger cities.

Many people are now starting to seriously consider returning to the small community living ideas where you know your neighbours and life doesn't seem so hectic.

The terms community, village and hamlet are generally interchangeable as they mean much the same thing - a form of human settlement where a group of people deliberately live nearby to each other for a particular purpose.

However, in this context, a "hamlet" refers to a cluster or group of 20 to 50 houses with facilities such as a hall, service station and general store.

A "village" means a larger grouping of houses (several hamlets) with the additional services such as a school, post office and other shops, while a "community" often means the people, rather than the houses, that live together in the same locality.

A community of people often share the same ideals and beliefs or hold similar ethics.

These ideals can be common religious beliefs, ethical beliefs or some other beliefs and values. For example, the term "ecovillage" often describes a group of people who share a concern for the environment and who wish to live in a sustainable way.

The village infrastructure is designed so that it has little impact on the environment, and the lifestyle of those living on the site is one of "treading lightly on the earth".

This describes the true essence of a permaculture village. People with like-minds, with commonly-held beliefs or aspirations, coming together to form villages, need to hold similar goals, or the

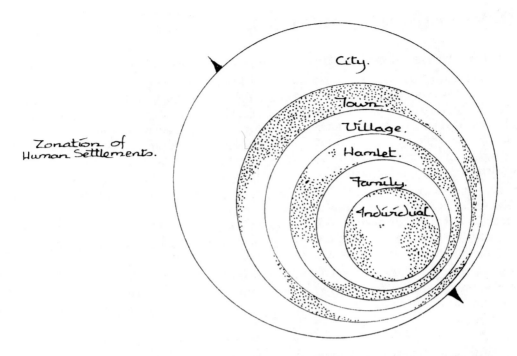

Figure 13.1 The concept of zoning can be symbolically applied to human settlements.

community will struggle to survive.

Living in a close community has many benefits. These include income generation on site (from a range of agricultural and financial activities), greater self-employment, less reliance on vehicles, machinery and equipment, the ability to pool labour (in building houses and other structures), greater provision of the basic human needs of food and shelter, reduced household operating costs, and on-site energy and electricity generation (thus less need to rely on state or private company services). In summary, the aims of any eco-village should be to:

- develop enterprises so that residents can earn a living on the site. This reduces transport costs and provides income and material goods for many residents.
- provide the basic life essentials of shelter, food and energy for each resident. This may mean that some residents earn income from growing food for others, generating power for the community, or building houses and other structures for individuals.
- provide services, such as education, and recreational facilities.

If more of these communities were set up then the incidence of social problems might decrease.

This, is turn, could mean huge savings in welfare and may reduce the need for counselling services and other government and private agencies.

Ecovillages are meant to only contain a small number of people. One hundred households, with a total population between two hundred to five hundred, would probably be the ideal.

At this number, LETS trading can be effective, and several people can work the land and derive income from businesses and activities on the property.

It is important to at least know each

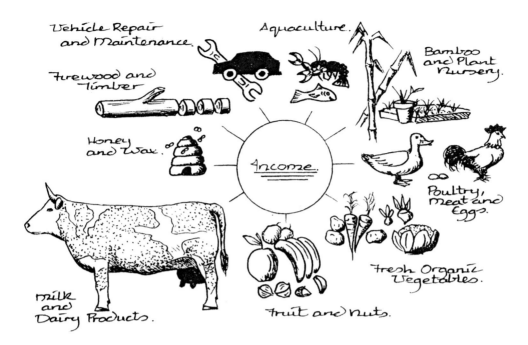

Figure 13.2 An example of income producing enterprises in a community.

other's names and be able to hold a conversation with another community member.

It is also possible to set up a school where funding from the regional or state education authority will often pay for a teacher's salary.

There are also enough people to maintain and develop the common areas and any reserves set aside by the community.

Most communities around the world are usually small, with perhaps less than one hundred people being the most common population size.

It has been estimated that by the year 2030 two-thirds of the world's population will live in major cities.

Finally, ecovillages should link into other similar settlements in the bioregion, so that the unnecessary duplication of services is reduced, and trade and support can be encouraged.

A variation to community life is the co-housing villages set up in Europe (originally Denmark) in the late 1970's. Here, a group of individual family homes are built surrounding a central, common house so that facilities can be shared.

In this way, individuals retain their privacy by living in their own units, but can participate in community life by having meals and activities together. The central house might contain laundry and full kitchen facilities, making these obsolete in every home or unit and thus reducing overall building costs. Some of the different ways that communities can be set up are discussed in the pages that follow.

Design considerations for ecovillages

Bylaws and regulations

The development of villages needs considerable research and thought. Each state, county or country has rules and regulations about the type of housing developments they will approve.

> ### *Guidelines for land development*
> Here is a short checklist of the types of considerations and issues that have to be addressed when communities are designed or built.
> - will development enhance the quality of life for those who live and work on the property?
> - will biodiversity, existing vegetation cover and soil and water quality be protected or enhanced?
> - are watercourses being changed or new waterways and waterbodies developed? Will these watercourses and the surrounding catchment area be protected? Will recharge areas be protected and rehabilitated?
> - how are wastes being treated? Will the soil be contaminated by the possible unintentional dumping of household rubbish?
> - will the proposed land use practices be sustainable? Will the environment benefit from activities undertaken on the land?
> - will the community development be compatible with the general character and land use of adjoining properties?
> - will the aesthetic qualities of the property be retained, or can visual improvements be made?
> - what will be the impacts of providing services such as water, power and telephone on the land?
> - is there provision for minimising the impact of feral animals, fire, drought, flood, hail and other natural disasters on the property?
> - are resources well-managed so that they are not quickly depleted?
> - does land use provide a range of opportunities for human enterprises?
> - will development enhance and complement local landcare initiatives?
> - has there been observation and consideration of the natural drainage and flow of water on the property?
> - will the development foster community awareness and education?
> - is there potential for rehabilitation of wildlife corridors?

You need to be aware of these regulations well before you start looking for land or a group to share it with.

Furthermore, the development of any ecovillage needs to fit into the existing or proposed plans for the bioregion.

Local, state or federal government authorities can be contacted for information about the restrictions and conditions for site development such as:

- road types needed - wide, narrow, gravel, brick-paved or bitumen.

    Each of these has advantages, disadvantages, different costs and particular applications for access requirements.
- power, telephone and water services layout - underground or overhead. Overhead power supply is cheaper to install, but underground is more aesthetically pleasing and safer in areas that have frequent storms and high winds.
- number of dwellings permitted. This may depend on the zoning of the land.
- nature of the titles - multiple occupancy, strata title or other. Some

arrangements permit individuals to own their own land within the site, while other set-ups have the community owning all of the titles and individuals leasing their house area.

There are many other combinations of conditions, each having particular rules and regulations.

- sewage and waste disposal options. This is of increasing concern to local authorities, as they and the general public are aware of the potential pollution problems to underground and surface water supplies.
- drainage, earthworks, restrictions on dams and waterways. For example, you may have to obtain pumping rights to take water from existing creeks and rivers, or you may not be able to clear some areas of the site for orchards or a house site without written approval.

Finding the right piece of land for you does present problems. Some are grounded in legal constraints, as we have just seen, and others are related to availability, cost, size and the specific features you desire, such as aspect, sun-facing slopes, and abundance and quality of water.

There are many ways in which land for villages becomes available and here are a few:

- a property owner wants to better use land he or she already has.

  Maybe a farmer wants to develop an ecological village on one hundred or more acres (40 ha) of land on part of the property.

- a group who have common interests search for and buy a block of land suitable for their needs.

- a person has a property and invites others to buy a share so that they can jointly develop it.

Which path you travel down depends on your circumstances and the situations that arise. Sharing that journey with others reduces the burden and is generally more rewarding and fruitful.

Looking at options

Like any permaculture design, the formation of ecovillages involves a series of stages such as research and collating information about the site, the design of the infrastructure on the landscape, and developing guidelines for the implementation and management phases.

The actual design of the community also develops in a series of stages - generally in the order of:

- water harvesting.
- roads and services.
- dwellings and outbuildings.
- primary producing areas, such as orchards, nursery and livestock.
- reserves and communal land, including forested areas, walkways, trails and recreation (such as ovals).

Ecovillages need to have structures and plans to maximise water storage, set aside land for housing, orchards and garden areas, natural forest and crop areas, and sites for community shops, school, hall and roadways.

Elements, such as those just listed, placed in the design need to be connected, rather than isolated objects on their own.

Humans need to be connected too, and human settlements and communities are an integral part of the permaculture lifestyle.

The development of particular infrastructure such as earthworks, buildings, roads and installation of services needs to be funded.

Potential buyers or residents of the village may need to provide money for site development, which can be placed in a trust and used as various stages occur.

Either the group pools their resources and funds in this way, or a land developer provides the financial outlay and he/she takes their percentage and costs from the sale of titles. The developer may either have the necessary funds or they will raise the capital from potential investors or through loans.

The costs for some stages may be paid by means other than money. Contractors for earthworks, road construction or other major service supplies may choose to take equity in the property, as one or more house sites, rather than cash. They may choose to become residents themselves or sell the house site/s at some later date - provided that this is legal in your state or country.

Not all villages need to be located on vacant or rural land. Ecovillages can be set up in a city by utilising adjoining houses in a block, or the flats or units of a housing complex.

A community really focuses on the human component, rather than the structural and physical components.

Any village development on rural land, however, generally has the greatest start-up costs. It is better for a group of people who want to form a village to look for land alongside an existing village or where essential services are already established.

It is especially important that abundant, clean water is available and that enough water can be harvested, stored and recycled on the site.

Furthermore, if energy generation is to be considered, then supplies of wood or high pressure water may be important. The site might lend itself to solar and wind harvesting, as alternative or appropriate energies such as these have less impact on the environment. Some of the criteria that you could consider when purchasing a block or looking for a site were discussed in Chapter 7.

On-site waste disposal is an important consideration. Dry composting toilets for each household minimise water use. All greywater can be treated at each house site or, where appropriate, at one or more central treatment plants. These could comprise a series of settling tanks or ponds and reed beds - in either ponds or subsurface flow.

The community should take responsibility for its own wastes and recycle or treat these as much as possible. We have to

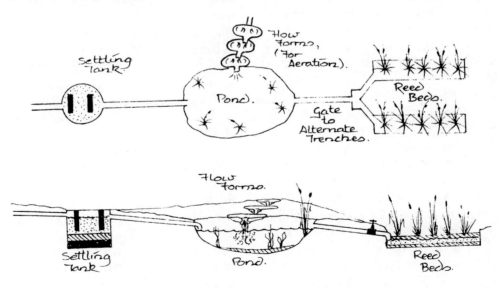

Figure 13.3 On-site waste disposal can be achieved for a community.

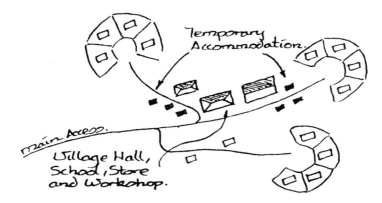

Figure 13.4 Larger villages might have a central hall, shops and workshops.

change the belief that the Earth has an endless capacity to absorb human wastes.

A village may refer to houses in a particular area, but it also contains the village centre where workshops, the school, stores, hall and other buildings are found. In particular, a community hall with meeting rooms and a kitchen is essential.

This may be used as a school; the community library; to run workshops and courses; general meetings of community members; the co-ordinating centre in cases of emergency, such as threats of fire; and for business operations, such as a coffee shop, fruit and vegetable co-op or a community deli.

Larger communities may want to build a large workshop where vehicle repairs, welding and wood turning can be carried out. The village design should advocate the minimal use of cars by designing for people and not machines.

Roads should be narrow or curved (even installing speed humps at the main road entry) to reduce the speed of traffic, and walkways and narrow pedestrian streets can be built.

Some vehicles are a necessity these days. Consider a bulk fuel supply for residents by having a community petrol bowser or petrol stored in tanks or drums at a central depot. This can mean savings to residents.

However, you usually have to obtain permission for fuel depots such as these from local and state authorities. Find out about the rules and regulations before you start planning.

A full range of tools such as lathes, posthole digger, chipper or mulcher, electric saws and so on, can be loaned to members through the community LETS system.

The community may also own a range of vehicles such as a backhoe, a front-end-loader (FEL) tractor with implements and attachments for soil conditioning, and a truck or van for carting large objects and goods. A bobcat is not needed as the slower backhoe and FEL tractor will do the same job.

The site design for the community, the implementation schedule and management plan should be made available to all residents.

There may also be legal requirements for certain management statements and property plans to be prepared.

In particular, the trustees of the land trust need to set up monitoring and reporting strategies on the development of the site as time passes.

Changes to the design, and other initiatives, should be recorded so that future monitoring and management can be effective.

Existing residents and intending buyers should be informed of any changes to the design and, wherever possible, actively involved in the review process.

Housing structures

A range of different types of dwellings is appropriate for villages. Most would be individual family homes, but the community may wish to have single or double bedroom units for temporary accommodation for visitors or workers (either contractors or employees). These units could also be used for emergency accommodation in times of family crisis.

Furthermore, elderly people may prefer small units, so a selection of housing types may be required to meet all of the needs of future residents.

The actual style of the house and its building materials should be left up to the individual.

However, it would be hoped that a community would insist that all houses have solar access, screens for privacy, and be able to utilise gravity for water movement - from greywater, roof run-off and so on, down towards garden areas or storage tanks.

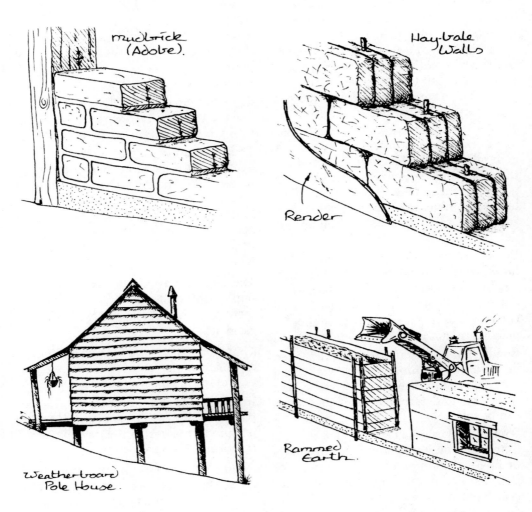

Figure 13.5 Many building materials can be used for house construction.

The community may also want to set standards for house design to avoid adverse visual impact, poor construction and the use of inappropriate building materials. Most houses can even have their own rainwater tank, thus reducing the burden on the community supply. This is not that critical in countries that have a year-round rainfall.

When allocating the number of titles or house sites for a particular development, about two-thirds to three-quarters, or more, can be sold to community members. The community may want the titles of the other quarter to a third to be kept, in trust, for use by the community to subsidise low-income families or to attract essential recruits who can offer services such as medical, administration, education, engineering and construction skills.

Part of the design process is to choose and mark any potential house sites. Clients can then pick which one they want.

It should be clear, and accepted by all community members, that they are not allowed to choose their own site in the bush or on top of hills or on the valley floor.

However, they may be able to select their actual site within the building envelope, as this will depend on topographical and other features of the site.

Houses should never be sited on ridgeways, as these sites are exposed to extremes of heat and cold, high winds and greater risk of erosion and ground movement.

Figure 13.6 One access road along the contour is a cost-effective strategy.

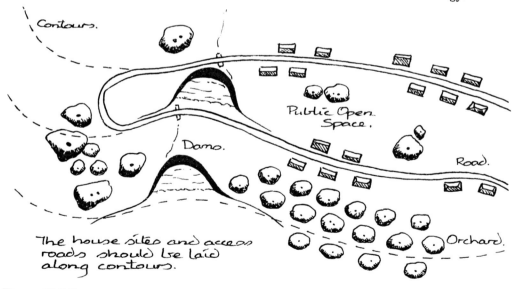

Figure 13.7 Sometimes houses can be placed along a few contour lines, so that more of a hillside or slope can be utilised.

Ideally, all house sites are placed in practical, warmer areas of the hillside. If a client wants a view, they can walk up the hill to an observation post where they can sit and view the surrounding countryside at their leisure.

Building houses in the thermal belt on a hillside makes sense. All services can be supplied along one main road on the contour as it curves around the hills.

This is a single service system - power, water, gas, telephone line and access road all along the same path, as shown in Figure 13.6.

Earthworks and contractors' fees will be minimised if the job is easy and simple to do, and can be done in a short time.

The arrangement of houses varies from community to community. Small clusters of about half-a-dozen houses (an enclave) is a good human settlement strategy, as there is some privacy as well as allowing for social interaction.

You can even have clusters as themes, such as art and craft, plants and propagation or wilderness conservation. People can then choose which group has similar interests to themselves and live in that area.

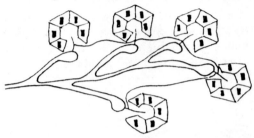

Figure 13.8 Clusters of houses (an enclave) is a good design strategy. There is a balance between privacy and community living and social interaction.

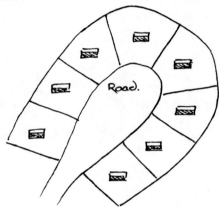

Figure 13.10 Another alternative is to have houses around a central access road.

Social and legal structures of human settlements

Early in the formative stage, prior to any development of the village, various aims, objectives and general statements about the ethics of the group should be developed.

This mission or ethic statement should be a broad, realistic goal that usually does not promote a particular belief system and which is universally accepted by all members of the community.

Often, the statement reflects a stewardship focus, such as "care of the Earth".

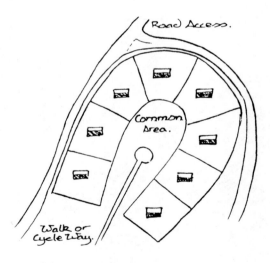

Figure 13.9 Design for people. Here, houses are grouped around a common area, with cycle paths and walkways to link small hamlets. The access road is positioned at the back of the houses.

Obviously, religious groups will want to specify the beliefs and values they wish all members to uphold, and this becomes their creed.

The ethics statement influences the direction taken in the development of any financial, legal and physical structures in human settlements.

Several trusts may need to be set up to administer these aspects of the community. For example, one trust may be needed for the sale of land and the general management of common land, including the future development of community buildings.

This would be a land trust that is shared equally by all members of the community or is used by particular residents for some purpose.

Legal structures are needed so that common land held by a trust is managed by a company (as trustee) with various elected residents as company directors.

This arrangement will then allow residents to be able to lease parcels of land for some small sum of money or undertake its use with an agreed-to arrangement.

For example, it might be agreed that ten percent of profits from an enterprise be returned to the community as community funds.

A second trust could handle all business trading operations. The business, money-earning activities from land use and other community enterprises should be managed by a trustee for this type of trust.

This trust needs continual income for the maintenance of roads, fences, water supply and other services, as well as capital expenditure for machinery and equipment.

It can generate income from the lease or hire of equipment, tools or land, or set annual charges (as a maintenance fee) that all households pay.

You can find out about trusts and how they operate from accountants and lawyers. These people will complete and submit the necessary paperwork to register a trust and trustee companies.

Trusts do have some rules, such as an annual meeting of trustees, but these are minimal and not overbearing.

It is possible for large communities to generate jobs for many villagers on the site - ranging from firewood collection to supplies of fruit and vegetables and meat, milk and milk products, herbs, nursery plants, building materials such as timber and mudbricks, and so on.

Many people should be able to earn some income from enterprises on the site.

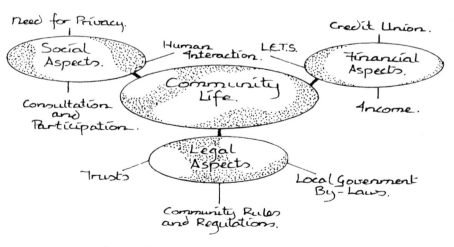

Figure 13.11 Some aspects of community life.

For example, a dairy processing centre is one option. One family could operate a dairy for goats or cows and sell milk and milk products such as yoghurt, cream, butter and cheese to villagers.

This is another example of keeping money inside the community boundaries and letting it circulate only within the community.

A number of small villages in a region can form a trading co-operative. This opens up other opportunities for trade, manufacturing and industry.

Scattered ecovillages themselves need to network with other villages to further pool resources.

Villages should encourage the settlement of a broad range of people who will provide natural diversity wthin the group.

Individuals or families, with a wide range of incomes, should be able to join the community. They may be able to "pay for their share" by sweat equity or some other satisfactory arrangement.

This means that instead of paying money for their share in the property, they may be able to physically work for the community in some way, from administration tasks to building houses, to maintaining common food-producing areas, to planting trees to rehabilitate degraded areas.

One option for the community is to set up a credit union as a financial body to make low-interest loans available to residents, so that everyone is able to have the same opportunities of living and working in the community.

Furthermore, the community may want to recruit particular individuals who they need for further development and expansion. Providing house sites, shares or other incentives will entice prospective community villagers.

The actual sale of land units or titles, held by a group of people, may have to be undertaken by a licensed real estate agent, depending on the laws and regulations in the state or country. Individuals can usually sell their own land. The necessary information and the offer and acceptance forms are readily available from newsagents and some stationery suppliers.

The other main aspect of community life is the community bylaws. Each community group needs to get together to decide the 'house rules', such as whether pets can be kept, and, if so, which ones are suitable, which ones are not, how many, which areas are restricted to pets and which allow access.

This should not be an exhaustive list of restrictions, but rather a short list of ten or so basic rules to which everybody agrees.

Various community members should be responsible for different aspects of community life. It is a good idea to give responsibility to different individuals, so that many residents are involved in the day-to-day running operations.

It might be that a small group of residents becomes responsible for administrating and formulating these bylaws, and they might develop regulations about music and noise curfews, or about how people can buy into or out of the community. For example, the arrangement for buying and selling housing sites can vary.

In some villages, residents can live in houses they build, but they cannot sell for profit.

If they leave, the ownership of the house reverts back to the community. Some money can be paid by the community as compensation.

In other communities, houses and a small area of land surrounding the house site are privately owned. Owners can sell their houses for whatever price they wish.

In some communities the new buyers have to be approved by the group, in other communities there are no such restrictions.

Permaculture communities, I believe,

should be developed in such a way that those who live on the property are all like-minded, and only those who adhere to the community ethics should be encouraged to settle in the community.

Good management skills are essential for the success of any community.

Administrators have to have a clear vision for the future potential and development of the community, including promoting responsible land use practices, having a good knowledge of financial dealings and constraints, and being able to make decisions that will benefit both individuals and all members of the community.

My notes

Things I need to find out

14 Appropriate technology

Appropriate technology is the use and promotion of machines, techniques and equipment that are environmentally friendly. It may be technology that harvests renewable energy sources such as the sun or wind, has low energy requirements itself or does not produce waste and pollution. Appropriate technology includes the use of alternative energy sources, such as biogas and solar energy, and most often can be easily built or developed. Only some aspects of appropriate technology are discussed here: those related to producing or using electricity or power, pumping water and cooking food.

Generating power

Photovoltaic cells

A photovoltaic panel or module is composed of many individual solar cells that are electrically joined together. Each solar cell only produces about 0.6 volt, so many are needed to produce sufficient voltage to recharge a battery or run an appliance such as a solar pump.

Solar panels come in a range of sizes with the most cost-effective being about the 40 to 90 watt (W) range. An eighty watt panel will produce up to five amperes of current when charging a 12 V battery. This is enough to trickle charge the battery during the day.

The solar panel may have a blocking diode in the output box so that current cannot flow from the battery back into the panel. A diode is like a one way valve which only lets electricity flow in one direction.

Polycrystalline panels require this diode because of current leakage, whereas monocrystalline panels usually do not.

You can monitor the rate of charge and the panel performance by wiring one ammeter into the circuit. Ammeters can be placed to measure both the current flowing into the battery bank and that flowing out through the circuit, as shown in the following diagram. A regulator is also placed in the circuit to stop the battery from overcharging.

While some appliances and motors can be run directly off a solar panel, most applications use a battery to store the electricity until it is needed. Batteries convert direct current (DC) into chemical energy which is stored on the battery plates. In effect, batteries use direct current to change the chemical state of the battery plates. This chemical reaction can be reversed to produce an electrical current.

Most house appliances rely on alternating current (AC) which is electricity flow that changes direction (alternates) about fifty times each second. To use the low voltage DC power to run high voltage AC appliances we need an inverter. Inverters basically convert DC into AC. Most have both 12 V and 24 V input, so various combinations of battery voltage can be used.

A full range of inverters are available and you should have no trouble finding one to suit your budget and requirements. Cheap inverters, however, may not produce a proper sine wave, so you may experience

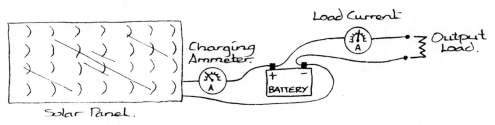

Figure 14.1 Ammeters allow monitoring of the electrical system.

background noise - a humming when appliances, stereos or amplifiers are used. You usually find that appliances will not work as well if you use an inverter which does not produce a true sine wave.

Always buy or recommend an inverter which produces greater power than that actually required as sometimes they are over-rated by the manufacturers; they can be slightly inefficient and they may only supply the stated wattage output for a short time, and cannot sustain continuous use at the higher output. Inverters should specify a continuous, intermittent and a surge rating. Many appliances, in particular motor-driven appliances, will require a start-up surge significantly higher than their continuous power consumption.

Solar panels are most effective when they are facing the sun. A manual or electronic tracker can be used to move the panel or panels so that it faces the sun throughout the day. Trackers follow the sun on one axis - east to west - and they are most beneficial in the early morning or late afternoon and in summer with the longer daylight hours. Trackers can squeeze up to 50% extra power during these times, depending on latitude and climatic conditions. Trackers aren't that effective in winter as the sun's arc is less than that in summer.

Manual trackers, where you physically move the panels during the day yourself, are only suitable for those people who can remember to do this task every now and again. You may have to move the panels four or five times a day to increase the effectiveness of energy production.

Solar panels alone are limited, by the availability of clear sunny days, in supplying the energy needs of a household. Most often a combination of energy harvesting strategies are used, in what is called a hybrid system. Here, solar panels may be used in conjunction either with a wind generator, diesel generator or hydro-electric system to produce power day and night and season to season. The principle of a hybrid system is that when one energy harvesting source is non-operational, another may be sufficient to keep the batteries charged or motors running.

Wind generators

The rotational motion of the blades of a windmill can be directly used to pump water or generate electricity. Wind generated power usually costs less to produce than solar panels if a windy site is available.

Solar panels do not get cheaper as the array gets larger. However, wind generators can become cheaper as they get bigger, making the unit energy production costs lower. Furthermore, wind generators may be up to ten times more cost-effective than solar cells. This means that

Figure 14.2 A solar array needs to be seasonally adjusted for the greatest current production.

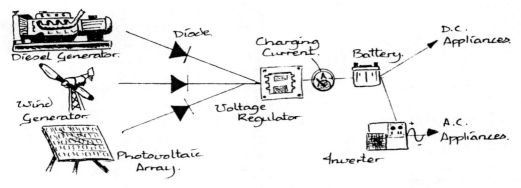

Figure 14.3 A hybrid system combines several energy harvesting strategies.

they proportionally produce more electricity and lose less energy in the system.

Efficiency is the ratio of the energy output compared to the energy input. For example, a car battery (accumulator) is about 80% efficient in converting chemical energy into electrical energy, whereas a solar cell is only about 10 to 15% efficient in changing light into electricity.

The amount of electricity produced by wind generators depends on factors such as the wind speed, size and number of propeller blades, and the degree of turbulence or variations in wind speed. Generally, wind speed increases dramatically with the height above the ground, so towers up to twenty metres high are used.

Wind towers need to be placed reasonably close to houses (less than 100 m if it is a DC turbine) because the further away it is, the greater will be the voltage drop, size of wiring and general installation costs. High voltage AC turbines can be placed further away from house sites. The high voltage output of some turbines can help overcome voltage drop and significantly reduce the cost of cable.

The tower should be at least 8 m above any obstructions for at least 150 m in any direction. This minimises the effects of air turbulence as wind passes over trees and buildings. Noise from wind turbines can be heard on their downwind side.

Altitude also plays an important part in calculating the size of the generator. Wind generators are rated for particular ranges of wind speeds. As you travel higher above sea level, the density of air changes and thus higher wind speeds are required for the same output.

Wind towers also need to be earthed

Figure 14.4 It is important to place wind generators in areas not subject to turbulence.

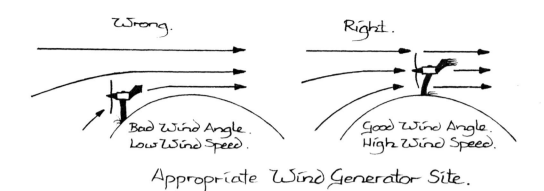

Figure 14.5 The siting of the wind generator is crucial to its success in supplying energy.

against lightning and guyed with cable, just in case of storms and very strong winds. It is advisable to monitor a potential wind site for at least one year before installing the tower and generator.

Use an anemometer to measure daily wind speeds to assess the viability of a site. You can usually buy a range of digital and hand-held anemometers which are accurate enough to assess the site.

Wind generators for domestic use are not that costly or large. However, if you wanted to supply the power requirements for a community the wind generator would be expensive, large and noisy.

They are ineffective at very low wind speeds, and at higher speeds (during severe storms, cyclones and hurricanes) damage can occur.

However, wind generators normally have a feathering device to protect them against high winds.

The tower is placed several hundred metres away, as the larger generators make a continuous hum which can be annoying at times.

You want to have the wind generator in the windiest site - but not your house.

Hydro-electric systems

Small hydro-electric systems will be much cheaper than the equivalent power production from solar cells if a good hydro site is available. If sufficient water flows all-year-round, power is generated day and night. Solar panels only produce current when the sun is shining and are just not suitable in countries or climates where the number of blue sky days is low.

The type of hydro-system you may need depends on the amount of water flow and head of pressure available.

The "head" is the vertical height that the water drops as it flows from the inlet pipe to the generator, as shown in Figure 14.7. Power output from a hydro-system is a product of head and flow, so with less head you will require greater volume of water (or increased flow) to achieve the desired wattage output.

Thus, when evaluating the use of a hydro-system you need to establish both the head and water flow. Greater head permits less volume and greater volume reduces the head requirement.

Hydro-electric systems complement photovoltaics as there is usually an abundance of water in the winter months or wet season but sunshine is scarce.

Pelton wheels, as shown in Figure 14.8, are examples of small hydro-electric power stations. They are designed to work best with low volume, but higher pressure, water sources. Pelton wheels generate electricity when a fine jet spray of water is directed into a series of cup-like buckets which cause the wheel to turn. The wheel is attached to a generating device and electricity is produced.

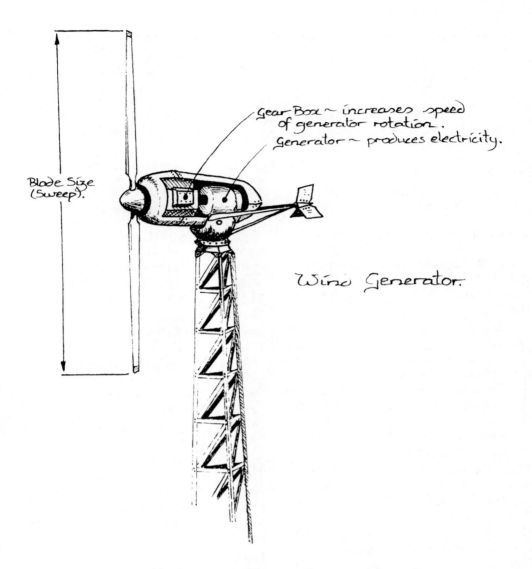

Figure 14.6 Wind generators are cost-effective in many areas.

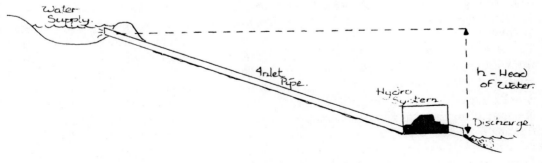

Figure 14.7 The head of water is the vertical height of the water column.

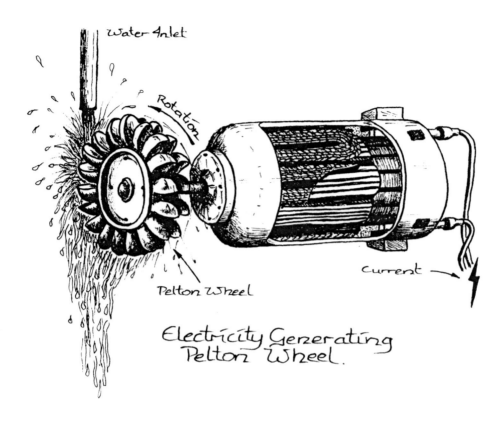

Figure 14.8 Pelton wheels generate electricity when a jet of water causes the wheel to turn.

Electric fencing

Electric fencing is cheap to build, generally reliable and can keep out predators and feral animals as well as containing stock and domestic animals.

You can either buy build-your-own kits from electrical suppliers or buy cheap commercial varieties from a range of farm supply stores.

Some require a motor vehicle coil and most need a 12 V lead acid (accumulator) battery. In this type of system, a small solar panel can be used to charge the battery.

You can also get mains power energisers, but these are generally dearer and they require a main power supply to run the equipment.

This is how it works. The energiser sends out a pulse of low current, high voltage electricity along the wire. When an animal touches the wire, the electricity can pass through the animal's body to the earth. The animal receives a high voltage shock for a split second, enough to make the animal wary of the fence in future.

There are two common electric fence systems. The simplest is a single wire or thread that is earthed at the control box.

The wire must be insulated from any other metal conductors or short circuiting will occur.

The second method uses more wire in an alternating pattern of "live" and "earth" wires. An animal will receive a shock when it simultaneously touches any live and earth wires. Nothing occurs if the earth wire alone is touched.

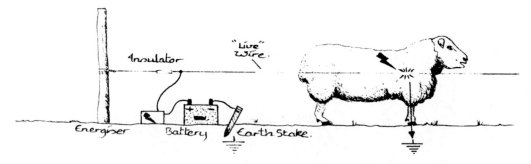

Figure 14.9 When an animal touches the wire electricity passes through its body to the ground. The circuit is complete.

Pumping water

Pumping water requires a fair amount of power. Water is heavy stuff. It weighs one kilogram a litre (over 8 pounds per gallon) and considerable power is often required to lift it great distances.

It would be great if we could all live on a hillside with a fresh mountain stream at the top so that water could be directed by gravity towards the house. This is the exception, not the rule. Few people have this luxury, so water has to be pumped and stored.

For garden sprinklers to function effectively, or for you to have reasonable pressure at the kitchen tap, we have to have about fifteen to twenty metres of head. We can get by with ten to fifteen metres but the pressure obviously is reduced and just trickles out of the garden hose.

You shouldn't recommend a head of water any less than twenty metres if you want crop and tree irrigation by gravity alone.

It may be all right for the client to have a weak supply to the kitchen, but there just won't be enough pressure to irrigate orchards and food crops.

You may be able to get a booster pump that will increase the pressure from the storage tank to the house. Alternatively, using larger diameter pipe to increase the flow of water achieves the same result as using higher water pressure.

Batteries can be used to pump water by

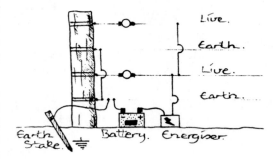

Figure 14.10 This system is more reliable as more wires are used and earthing occurs easier.

supplying the necessary energy to run a DC pump.

It is also possible to run the pump directly off a solar module or wind generator if the conditions are right.

However, small, cheap electronic devices, called maximisers, can be built or bought to amplify the current output from solar panels or wind generators when the sun is partly covered or wind speed is low.

Maximisers examine the wattage coming in and what is required by the battery or pump and change the output accordingly.

Maximum power can be obtained from a solar panel or wind turbine by continuously adjusting the voltage to suit the application.

This, in turn, produces the maximum current so that the power (watts) remains the same.

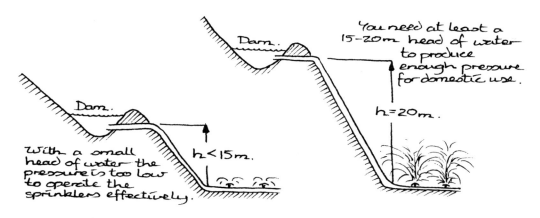

Figure 14.11 To irrigate crops and trees by gravity you need at least 20 metres of head.

Solar pump

As we have just discussed, pumps can run off a solar panel directly. Modern electronics is such that the low current produced, even during cloudy days, can be increased (maximised) by electronic circuitry and the pump will continue to work most of the day, lifting and shifting water to where it is used or can be stored for later use.

Many AC submersible pumps consume three to four times more power than a DC powered pump to move the same amount of water. Generally, AC submersible motors are harder to start, are not designed to be energy-efficient nor engineered with the quarterly power bill in mind.

DC powered pumps that are run directly off a solar panel are the cheapest and easiest solution to water supply problems. As long as the sun shines the pump will work - even if it is very slow at times. Remember that the total volume of water pumped by a submersible pump decreases with the total lift from the bore or dam. The deeper the pump is, the less water that can be pumped each day - for that size of pump.

Hydraulic ram

Hydraulic rams use the water itself as the power source. Water flows down a drive pipe to the ram, a valve closes and the hydraulic hammer effect forces some water into a delivery pipe or small tank. Most of the water is expelled from the ram when that same valve opens. Essentially, a large amount of water falls a short distance, causing a small amount of water to be lifted a greater distance.

While hydraulic rams can only effectively transfer 10 to 20% of the water that enters, they can lift this water twenty times

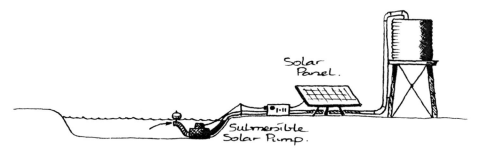

Figure 14.12 Submersible solar pumps are very reliable.

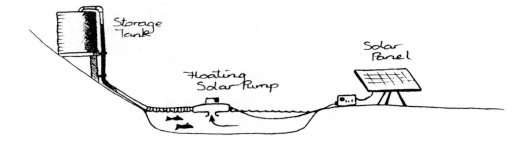

Figure 14.13 A floating solar pump is another option.

the incoming (drive) head. For example, a 3 m supply head can be pumped to a tank or dam 60 m up the hillside.

Some hydraulic rams can be placed directly in the water of a stream, while others are placed on dry land and the inlet pipe carries water from the stream to the pump.

Cooking devices

Solar oven

Appropriate technology can be used to make a range of cooking devices. Three common ones are described here. Solar ovens are usually made by painting the inner lining of a cardboard or wooden box with matt black paint. This increases heat absorption and re-radiation. About 50 mm of wool, fibreglass or some other type of insulation is placed between the inner and outer box walls. A clear glass lid, preferably double glazed, is sealed to keep

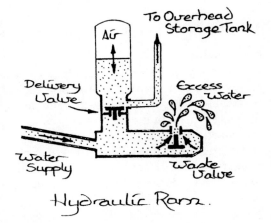

Figure 14.14 A hydraulic ram uses water as a source of power to drive it.

the heat in.

The solar oven is placed in direct sunlight and moveable side flaps and lid flap (all coated with aluminium foil) are adjusted so that sunlight is directed into the oven.

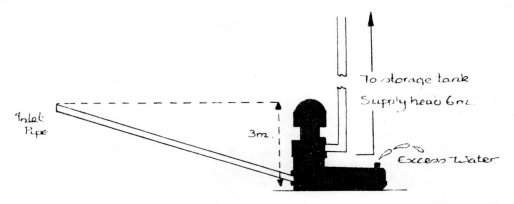

Figure 14.15 Hydraulic rams work day and night to pump water uphill to a storage tank.

Solar ovens will heat up to well over 150°C, which is enough to cook most things - even if it does take a little longer than conventional wood, gas or electric ovens.

Solar food dryer

A solar food dryer is an effective way to dry excess fruits and some vegetables. Sunlight is absorbed by the black rear surface and re-radiates this as heat.

The air slowly becomes hotter inside the dryer and rises upwards and vents out at the top. As the hot air rises, cool air is drawn in at the bottom and this, in turn, heats up as well.

The fruit slices are placed in trays at the top. The hot air passes through the trays and dries the fruit, usually within a few hours.

Haybox cooker

A haybox cooker doesn't have to be made out of hay, although hay is a very good insulator. The cooker doesn't supply any energy to cook the food either. It simply allows the food to continue to simmer long

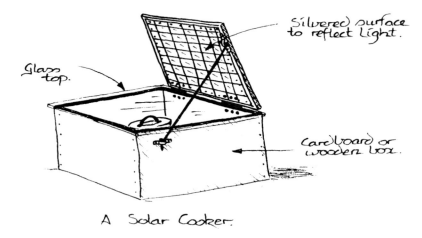

Figure 14.16 A solar oven uses free energy to cook food.

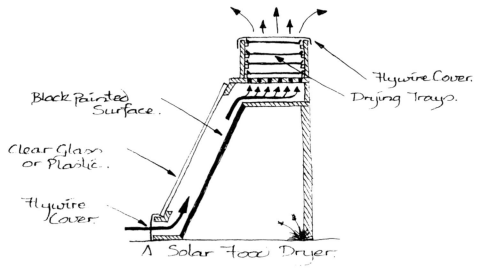

Figure 14.17 A solar food dryer allows you to preserve some foodstuffs for later use.

after the food is removed from a stove or burner.

The haybox cooker is an insulated box that helps to contain the heat of the actual cooking vessel inside it.

A stew or other dish is cooked for a short time, usually until it boils for a few minutes, and is placed inside the haybox cooker.

A lid is tightly positioned and the cooker is left for as long as you need. The heat from the cooked meal is prevented from escaping and this energy can continue to cook and simmer the food for several hours.

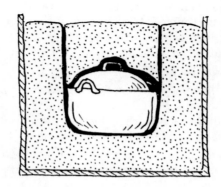

Figure 14.18 A haybox cooker keeps the heat in so that food continues to cook slowly.

My notes

Things I need to find out

15 Glossary

Active solar: a system harvesting and using the sun's energy, but where some energy is also expended in moving air or water (usually by pumps or fans) through the system.

Afforestation: planting trees in areas currently devoid of trees.

Agroforestry: farmland where trees are grown for timber or other uses, as well as crops or pastures.

Alley cropping: rows of trees separating bands of crops.

Alternating current (AC): electricity which reverses its direction at a constant rate. Nominally 50 hertz (cycles per second).

Anaerobic: respiration without oxygen. Various noxious gases, such as methane and ammonia, as well as carbon dioxide, may be produced.

Animal husbandry: the practices involved in the management and welfare of raising farm animals.

Annual plants: plants that complete their life cycle within one year.

Appropriate technology: the promotion and use of low-energy, low-maintenance, environmentally-friendly equipment and materials.

Aquaculture: food production using water as the medium.

Arable land: land that can be cultivated and used for crops or pasture.

Arid: areas where the rate of transpiration and evaporation exceeds precipitation.

Aspect: direction or orientation of the slope of land. e.g. facing north or south.

Azimuth: angle between north and the direction of the sun, on the horizontal plane.

Biennial plants: plants that take about two years to grow and produce seeds.

Biodiversity: the diverse range and variety of life in a region.

Biological control: the use of natural predators to combat pests.

Biomass: the weight or mass of the living components of the ecosystem.

Biota: living organisms.

Bunyip level: common name for a water level which can be used to establish marks or points at the same level.

Carbon-nitrogen ratio: the ratio of carbon atoms to nitrogen atoms in organic matter. The decomposition rate of compost increases with a CN ratio of 20:1 or less.

Catchment area: the land area that collects water for a particular stream or waterway.

Cation: positive charged atom, usually of a metal or hydrogen.

Chicken (animal) tractor: a portable, fenced pen structure which holds small numbers of poultry or other animals.

Chinampa: series of channels of water between fingers of land. A very productive aquaculture system.

Chisel ploughing: tillage operation where soil is shattered but not inverted.

Climax species: dominant plant species in an ecosystem.

Cold frame: a box, with a glass lid, which acts as a mini-hothouse. Used to germinate seeds in the colder months.

Combustion: the burning of organic matter, which releases water, ash, gases and smoke into the atmosphere.

Companion plants: plants, such as herbs, which are beneficial to others nearby. For example, they may repel pests, disguise a food crop or exude beneficial chemicals into the soil.

Compost: a mixture of decomposing plant and animal material.

Community: a group (of any size) of like-minded people who live, and sometimes work, together on the same property.

Condensation: process of water vapour changing to a liquid, such as dew on the surface of leaves or the ground.

Conservation: looking after our environment, which includes natural forest areas, soil and waterways.

Contour: imaginary line with all points at the same altitude or level.

Contour bank: trench and bank dug along the contour which catches water and prevents erosion.

Convection: transfer of heat by the natural flow or circulation of air or water. Warmer air or water tends to expand and rise carrying heat with it as it moves.

Coppicing: chopping trees down at stump level, where they will regrow.

Cover crop: plants grown to reduce erosion, trap nutrients and protect the soil.

Crop rotation: growing a succession of different crops on the same area of land.

Current: the flow of electricity. Measured in amperes (amps).

Deciduous: tree that loses its leaves during winter and becomes dormant.

Diode: small electronic device which only allows electricity (current) to pass through it in one direction.

Direct current (DC): electric current that flows in only one direction, from negative to positive.

Direct seeding: seeds are placed directly into the ground, often by a machine towed behind a tractor.

Diversion drain: drain which moves water off the land into dams or other waterways.

Drench: a chemical given orally to animals, usually to control internal parasites.

Dumpy level: a telescopic instrument which allows vertical height and horizontal distance to be calculated. Used to mark out contour lines and determine slopes for drains.

Dynamic accumulators: plants which take in and accumulate high levels of particular minerals from the soil.

Ecology: study of the interaction between organisms and their environment.

Ecosystem: a system in which matter and energy flow. It consists of a community of plants and animals and the surrounding environment.

Ecovillage: a community of people who live in buildings and surroundings which are designed to be energy efficient and have low impact on the environment.

Edge: the boundary between two different ecosystems, such as where land meets water.

Element: anything which is used in a permaculture design. Elements include plants, animals, humans, structures and environmental components such as rocks and water.

Environment: the living and non-living surroundings of an organism. Composed of biotic factors, such as predators and parasites, and abiotic (non-living) factors such as water, air and temperature.

Erosion: the movement of soil material from one area to another. Caused by wind, water or gravity.

Espalier: trees, usually fruit, which are pruned and trained to grow flat against a wall or on a trellis structure.

Evergreen: plant that keeps its foliage all year round.

Exchange capacity (of soil): a measure of the ability of the soil to exchange mineral ions with a plant.

Fertiliser: any substance, such as manure or chemicals, added to the soil to improve fertility.

Fodder crop: a crop grown to be cut and fed to animals.

Forage crop: a crop grown to be grazed by animals.

Germination: process where seeds start to sprout and grow.

Greenhouse: (see hothouse).

Green manure: plants grown and then slashed or cut and turned into the soil to provide additional nutrients to the soil.

Greywater: all waste water, except toilet water, from sources such as the kitchen, bathroom and laundry.

Guild: an assembly of elements such that they are mutually beneficial. Can consist of any combination of plants, animals and physical factors such as rocks, soil and water.

Hamlet: small group of houses in a community.

Haybox cooker: insulated box which retains heat from a cooking pot so that the food can continue to simmer for some time after being removed from a stove or fire.

Head (water): the vertical difference, measured in metres, between the top and bottom of a water column.

Hothouse: covered structure, with plastic or glass, which keeps the internal temperature higher than the surrounding atmosphere. Used to germinate seeds and protect plants during the cold winter months. Also known as a greenhouse.

Humidity: amount of water vapour in the atmosphere.

Hydraulic ram: machine that uses water to provide the energy needed to pump some of that water to another location or storage area.

Hydro-electric system: a system or machine that generates electricity by harnessing flowing water, causing a turbine to turn.

Insolation: a measure of the amount of solar radiation over a particular area.

Ions: charged atoms.

Integrated pest management: the use of several strategies to control pests, such as companion planting, pest traps and crop rotation.

Interceptor bank: a trench and bank cut into the clay layer so that surface or subsurface water is able to be collected and prevented from further downward flow.

Inverter: a machine which converts DC into AC so that conventional, domestic electrical appliances can be used.

Keyhole bed: a garden bed surrounding a keyhole shape (which is normally the path you step into to access the plants).

Keypoint: position on a slope where curvature of the land changes from convex to concave. Located by finding a contour that suddenly increases in width when compared to contour lines either side of it. Usually is found at the start of a valley.

Land breeze: breeze produced by differences between air pressure on the land and water. Air flows from the land mass towards the ocean or sea.

Land degradation: the reduction in the productive capability of the land due to soil erosion, salinity or other factors.

Leaching: movement of soluble nutrients from the top soil downwards into the water table.

Legume: plant which can manufacture its own nitrogen with the help of bacteria living in the plant roots. Examples include peas, clover and lucerne.

LETS: Local Exchange Trading System. A community-based scheme trading in "green dollars" for goods and/or services.

Microclimate: local climate differences which occur in a confined or small area. For example, it might refer to the small temperature differences either side of a plant where one side is exposed to direct sunlight and the other side is shaded, and therefore cooler.

Minimum tillage: ground is not cut up and ploughed as usual, but lightly loosened as seeds are planted.

Monoculture: the growing of only one type of crop on a particular area of land.

Mulch: natural or artificially applied protective covering to the soil. Material used as mulch includes chipped tree prunings, stone, plastic sheeting and straw.

Multi-graft fruit tree: a fruit tree that contains at least two different varieties of the same fruit grafted onto the tree.

Nurse tree: fast-growing shrub or tree which protects, usually from wind, sun or frost, another slower-growing, often commercial species.

Passive solar: the gain of solar energy without the use of additional energy expenditure. For example, a house able to be heated by sunlight being absorbed by the floor and walls.

Pasture: plants, usually grasses and legumes, which generally provide grazing stock with their nutrient requirements.

Pattern: the shape, geometry or form of an edge, including circles and spirals.

Pelton wheel: hydro-electric device which generates electricity as flowing water passes over an internal cupped wheel.

Perennial: long-lived plant.

Permaculture design: an integrated, holistic landscape practice. Usually physically represented by a detailed drawing of the potential development of a site. The design also considers the interaction and inter-relationships between organisms and their surroundings.

Permafrost: areas where the ground is permanently frozen.

Perspiration: water loss from the skin or surface of an animal.

Photovoltaic cells: solar cells which produce electricity (current) when sunlight shines on them.

Pioneer: plant which is first to invade and grow on cleared or degraded areas. Usually these plants are short-lived and fast growing.

Piezometer: inspection pipe in the ground which permits determination of the height of the water table and possible salt contamination.
For example, salty water exerts higher pressure than freshwater and thus rises higher in the pipe.

Ploughing: tillage operation where soil is shattered and partially inverted to kill and bury existing vegetation.

Polyculture: the growing of several different crops, plants or animals on a common area of land.

Precipitation: water from the atmosphere which returns to the Earth's surface, mainly as rain, hail or snow.

Prevailing wind: the most common direction that wind blows.

Radiation (heat): invisible heat energy given off by a hot object.

Reforestation: replanting native or indigenous species in areas which have been cleared.

Retrofitting: changing the structure of an existing house to make it more energy-efficient.

Riparian: the plants and ecology of the banks of a river, stream or other waterway.

Ripping: tillage of sub-surface soil without inverting soil layers. Tynes are dragged up to a metre below the surface of the ground, breaking up the hard-crust clay layer to permit better water and air penetration.

Rotational grazing: moving livestock from one paddock to another.

Salinity: the amount of salt in water or soil.

Sea breeze: cool breeze which blows, usually during the day, from the ocean or sea towards (and over) the hotter land.

Sector: region of a property that has particular external energies such as fire, wind, sun and water moving through the site. Sector planning allows us to control, direct or harness these energies.

Shadehouse: covered or partly covered structure which shades and protects plants during the hot summer months.

Sheet mulching: a technique of garden building where layers of mulch and compost are used to make garden beds on top of the ground.

Shelterbelt: grouping of trees which protect stock as well as acting as windbreaks.

Slope: the gradient of the land. Usually expressed as the angle of the ground to the horizontal plane.

Soil: the particles of the Earth's surface, which contain both inorganic and organic matter as well as living organisms.

Soil conditioning: changing the structure of soil by physical or chemical means.

Solar array: a group of solar panels electrically connected together.

Solar food dryer: structure which uses the sun's energy to heat air which, in turn, dries food slices or pieces.

Solar oven: insulated cardboard, wooden or metal box, with a clear glass lid, which uses the sun's energy to cook food.

Solar pergola: angled, bladed structure which allows winter sunlight to pass through into a room or house, but shades out summer sunlight.

Stacking: the placing and layering of as many plants as possible in a particular area. Some will be ground dwelling, some herbaceous and others taller shrubs and trees.

Stocking rate: the number of one type of livestock animal kept on a hectare of land.

Strip cropping: crops are grown in strips or bands to help reduce water and wind erosion of the soil. Trees are sometimes planted between the strips or rows.

Strip grazing: livestock grazing fenced-off sections of a paddock.

Succession: the slow progressive change in an ecosystem.

Suntrap: semi-circular structure, usually as a row of trees, which reflects sunlight and contains it, thus creating a warmer microclimate.

Sustainable agriculture: a system of agriculture which maintains productivity without detrimental effects on the ecosystem, over a long time.

Swale: ditch dug on a contour so that water can be held long enough for it to soak into the ground where it can be used by plants.

Terrace: land that has been stepped down a slope - levelled and stepped down to the next level and so on.

Theodolite: surveying instrument which measures the horizontal and vertical angles of landforms.

Tillage: practice of turning the soil. This often involves tractors and ploughs on farms, but also refers to garden forks and other implements being used in urban yards.

Topography: the type of terrain and landforms present, e.g. rock outcrops.

Transpiration: loss of water from the leaves of a plant.

Trellis: structure that allows climbing plants and vines to grow on it - either vertically upwards or horizontally across.

Tyne: a steel point on a tillage implement which disturbs the soil as it is dragged through it.

Voltage regulator: electronic device which produces a constant voltage output even though the input voltage may vary.

Water cycle: natural process where water, in its many forms, circulates between the oceans and other waterways, the atmosphere and the land.

Water table: the top of permanent ground water in the soil.

Windbreak: trees and shrubs, or structures, which are used to reduce wind speed and thus protect an area from damage or desiccation.

Wind generator: machine which produces electricity by the action of wind turning the blades of a turbine.

WISALTS: Whittington Interceptor Sustainable Agriculture Land Treatment System. A system of drains, cut into the clay layer, which are designed to capture water and direct it to other parts of a property.

Zone: imaginary regions around a house, or some other focus, indicating the location and placement of plants, animals and structures in a design.

16 Bibliography

About Permaculture

Books

Bell, G. (1992). *The Permaculture Way*. Thorsons. London.

Bell, G. (1994). *The Permaculture Garden*. Harper Collins. London.

Brown, D. (Ed.) (1989). *Western Permaculture Manual*. A publication of the Permaculture Association of W.A. Cornucopia Press. Perth.

Firth, J. (1996). *Permaculture - Dry Coastal Regions*. Yilgarn Traders. Geraldton.

Holmgren, D. (1992). *Permaculture in the Bush*. Nascimanere. Maleny, Qld.

Lindegger, M. and Tap, R. (Eds.) (1990). *The Best of Permaculture*. Nascimanere. Maleny, Qld.

Mars, R. and J. (1994). *Getting Started in Permaculture*. Candlelight Trust. Perth.

Mars, R. and Willis, R. (Eds.) (1996). *The Best of PAWA Volume 1. Selected Articles from the Newsletters of the Permaculture Association of Western Australia*. Candlelight Trust. Perth.

Mollison, B. (1994). *Introduction to Permaculture (2nd Ed.)*. Tagari Publications, Tyalgum.

Mollison, B. (1988). *Permaculture: a Designers' Manual*. Tagari Publications, Tyalgum.

Morrow, R. (1993). *Earth User's Guide to Permaculture*. Kangaroo Press. Kenthurst.

Watkins, D. (1993). *Urban Permaculture*. Permanent Publications. Clanfield (UK).

Whitefield, P. (1993). *Permaculture in a Nutshell*. Permanent Publications. Clanfield (UK).

Whitefield, P. (1996). *How to Make a Forest Garden*. Permanent Publications. Clanfield (UK).

Woodrow, L. (1996). *The Permaculture Home Garden*. Penguin Books. Melbourne.

Journals

Permaculture International Journal. PO Box 6039, South Lismore N.S.W. 2480. Four issues per annum.

The Permaculture Edge. PO Box 148, Inglewood, WA. 6052. Four issues for $16.

Permaculture Magazine. Permanent Publications. Little Hyden Lane, Clanfield, Hampshire PO8 ORU England.

About Organic Gardening, Herbs and Growing Vegetables

Books

Bennett, P. (1989). *Organic Gardening*. National Book Distributors. Sydney.

Bremness, L. (1988). *The Complete Book of Herbs*. Readers Digest Press. Melbourne.

Bremness, L. (Ed.) (1990). *Pocket Encyclopedia of Herbs*. Dorling Kindersley. London.

Conacher, J. (1980). *Pests, Predators and Pesticides*. Organic Growing Association W.A. Perth.

Cundall, P. *Organic Gardening*. Gardening Australia Collector Series No. 1. Federal Publishers. Sydney.

Deans, E. (1992). *Esther Dean's Gardening Book*. Angus and Robertson. Sydney.

Fanton, M. and J. (1993). *The Seed Saver's Handbook*. Seed Saver's Network. Byron Bay, N.S.W.

French, J. (1989). *Organic Control of Common Weeds*. Aird Books. Melbourne.

French, J. (1990). *Natural Control of Garden Pests*. Aird Books. Melbourne.

French, J. (1991). *Jackie French's Guide to Companion Planting in Australia and New Zealand*. Aird Books. Melbourne.

French, J. (1992). *The Wilderness Garden (Beyond Organic Gardening)*. Aird Books. Melbourne.

Kourik, R. (1986). *Designing and Maintaining Your Edible Garden Naturally*. Metamorphic Press. Santa Rosa.

Little, B. (1992). *Backyard Organic Gardening in Australia*. Sandpiper Press. Sydney.

Little, B. (1993). *Companion Planting in Australia (2nd Ed.)*. Sandpiper Press. Sydney.

Roads, M. (1989). *The Natural Magic of Mulch*. Greenhouse Publications. Elwood (Vic.).

Smith, K. (1993). *The Australian Organic Gardener's Handbook*. Lothian. Melbourne.

Wilkinson, J. (1989). *Herbs and Flowers in the Cottage Garden*. Inkata Press. Melbourne.

Newspaper
Acres Australia. PO Box 27, Eumundi. Qld. 4562. Six issues per annum.

About Appropriate Technology, Housing and Lifestyle
Books
Ballinger, J., Prasad, D. and Rudder, D. (1992). *Energy Efficient Australian Housing*. Australian Government Publishing Service. Canberra.

Energy Victoria. (1995). *Guidelines for Building an Energy Efficient Home*. Renewable Energy Authority. Melbourne.

French, J. (1992). *Backyard Self-Sufficiency*. Aird Books. Melbourne.

Gordon, S. (1987). *Complete Self-Sufficiency Handbook*. Doubleday. Sydney.

Harris, M. and Hutchinson, A. (Eds.) (1990). *Build Your Own Green Technology*. ATA Publications. Melbourne.

Harris, M. and Beaumont, L. (Eds.) (1993). *Green Technology House and Garden Book*. ATA Publications. Melbourne.

Paolino, S. (1992). *Living with the Climate*. Graphic Systems. Perth.

Pedals, P. (Ed.) (1993). *Energy From Nature (5th Ed.)*. Rainbow Power Company. Nimbin, N.S.W.

Schaeffer, J. (Ed.) (1992). *Alternative Energy Sourcebook*. Real Goods. Ukiah (USA).

Smith, K. and I., and Thomas, A. (Eds.) (1992). *The Australian Self-Sufficiency Handbook*. Viking O'Neil Publishers. Melbourne.

Journals and Magazines
Earth Garden. RMB 427, Trentham. Vic. 3458. Four issues per annum.

Grass Roots. Night Owl Publishers. PO Box 242, Euroa. Vic. 3666. Six issues per annum.

ReNew (formerly Soft Technology). Alternative Technology Association. 247 Flinders Lane, Melbourne. Vic. 3000. Four issues per annum.

17 Index

A

A-frame level   46
active hot water system   75
afforestation   122
agroforestry   121
aids to design   38
air panels, solar   82
alley cropping   121
alternating current   148
alternative energy sources   148
altitude, of sun   65
ammeters   148
animals, characteristics   124
animals, in permaculture   104
animals, rural   122
appropriate technology   148
aquaculture   14
aspect   26, 68
assessing a site   68
azimuth angle   65

B

balcony gardening   108
banana circle   100
bare-rooted trees   63
bees   105
bentonite   57, 64
bog garden   103, 134
Bunyip level   45
bylaws, community   146

C

catch crops   56
cation exchange capacity   52
chemical trap   90
chicken pen, rotating   106
chicken tractor   107
chickens   106
chill hours   25
chinampa system   17
chisel plough   55
choosing a property   67
circle garden bed   11
classification of soil   51

clay hardpan   95
climate, local   65
climax species   22
clusters, houses   144
co-housing   137
cold frame   114
community buildings   141
community bylaws   146
community gardens   130
community living   135
community mission statement   144
companion plants   57
compass, for drawing   50
compass, magnetic   44
compost   99, 133
compost bins   100
condensation   66
conditioning a pond   15
conditioning of soil   54
contour, definition   86
contour lines   36
contour map   86
cooking devices   156
coppicing   126
crop rotation   57

D

dam location   89
dams, keyline   87
dams, storage ratio   88
deciduous trees   75
deep litter poultry pen   107
design considerations   34
design, of houses   72
design process   32
design report   39
design steps   35
designing for catastrophe   28
dew   98
direct current   148
diversion drains   92
dolomite   57
drains   90
drains, diversion   92
drawing aids   50

dryland water harvesting   96
ducks   106
dumpy level   47

E

earth mound   27
earth-covered houses   73
earthworm farm   104
earthworms   55, 104
ecosystem   4
ecovillage   135
ecovillages, designing   137
edge effects   10
electric fence   125, 153
element   18
energy audit   130
espalier, fruit trees   111
ethics, permaculture   2
evergreen trees for climate control   75
exchange capacity   51
exchange capacity, cation   52

F

fire   28
flood irrigation   92
fodder trees   125
food dryer, solar   157
frogs   102
frost   24

G

garden paths   63
geese   106
generators, wind   149
germination of seeds   113
green manure crop   56, 57, 96
greenhouse   77
greywater   85
guidelines for designing schools   128
guidelines, land development   138
guild, rock   63
guilds   62
gypsum   57

H

hamlets   135
hanging baskets   110, 113
hardpan, clay   95

harvesting water   84
haybox cooker   157
head, of water   151
herb lawn   102
herb spiral   11
hot water systems   75
hothouse   77
house, construction materials   142
house design   72
house site, determination   69
houses, clusters   144
houses, earth-covered   73
hybrid power system   149
hydraulic ram   155
hydro-electric power   151

I

implementation, of design   41
indicators of soil   54
insolation   65
integrated pest management   56
integrating house and garden   75
interceptor banks   93, 95
inverter   148

K

keyhole beds   11
keyline cultivation   86, 88, 95
keyline dams   87
keypoint   87

L

land breeze   117
land clearing, problems   95
land development, guidelines   138
land titles   143
land trust   145
land use   68
land use, restrictions   69
LETS   136
level, "A" frame   46
level, Bunyip   45
level, dumpy   47
level, theodolite   47
light box   48
lime   57
limited gardening spaces   108
local climate   65

M

magnetic compass   44
Mandala keyhole garden   132
manure, animal   57
maximisers   154
meter, pH   45
meter, salinity   45
microclimate   23, 79, 101
minimum tillage   55
mulch   53, 98, 101

N

natural resources   68
non-wetting soil   64
nurse trees   121

O

organic fertiliser   57
organic produce   51

P

passive hot water system   75
passive solar house   72
paths   63, 132
paths, living   102
pattern, in design   10
pelton wheels   151
penetrometer   44
pergola   79, 110
pergola, solar   81
permaculture   1
permaculture design, rural   115
permaculture design, schools   127
permaculture design, urban   99
pest management, integrated   56
pH meter   44
photovoltaic panel   148
piezometer   95
pioneer species   5
plane table   47
planter box   109
plants, companion   57
plants, indicators of soil   54
plough, Wallace   55
plough, Yeoman   55
plucking beds   12
polyculture   15, 56

ponds   102, 134
ponds, bog garden   103
poultry   106
predators, in garden   58
property, choosing   67
protractor, for drawing   50
pumping water   154

R

rainwater tanks   84
raised garden beds   12
recharge areas   94
reed beds   85
reforestation   122
regulator   148
resources, site   68
retrofitting, houses   80
ripping, ground   90, 116
rock dust   57
rock guild   63
rotating chicken pen   106
rotation, of crops   57

S

salinity meter   45
scale ruler   48
school, design guidelines   128
school designs   127
schools, needs analysis   127
schools, recycling centres   130
sea breeze   117
sector plan   77
sectors   22
seed, germination   113
sheet-mulched garden beds   100
shelterbelts   117, 120, 121
silt trap   90
site analysis   40
site, assessment   68
slope   68
sod roof   73
soil   51
soil collapse   93
soil collapse, testing   94
soil conditioning   54
soil, non-wetting   64
soil treatments and amendments   56
solar air panels   82

solar chimney 74
solar food dryer 157
solar hot water systems 75
solar house, passive 72
solar oven 156
solar panel 148
solar pergola 81
solar pump 155
spreader drains 92
stacking 60
stacking, time 61
stock lock-up areas 123
storage ratio, of dams 88
succession 4
surveying equipment 45
sustainable land use 7, 115
swale 90, 96

T

tanks, rainwater 84
tape measure 43
terracing 84, 92
theodolite 47
tillage, minimum 55
time stacking 61
titles, land 143
tracker, for solar panel 149
transpiration 66
treating greywater 85
tree belts 121
trees, fodder 125
trees, for climate control 75
trellis 12, 79, 110
trusts, community 145

U

underground houses 73

V

vents, in houses 82
village living 135

W

Wallace plough 55
waste disposal, communities 139
waste disposal options 140
water cycle 66
water, harvesting 84

water harvesting, dryland 96
water "head" 151
water, pumping 154
water repellency 64
waterlogging 93, 95
wetland plants 85, 90
whole farm plan 115
wind generators 149
wind towers 150
windbreak, construction 117
windbreaks 116
window sills 108
WISALTS 93

Y

Yeoman's plough 55

Z

zenith 22
zones 19

Notes

"We can make forests into deserts or deserts into forests. It is our choice"
Bill Mollison, 1994

Printed in the United States
63839LVS00001B/1-14